# Advances in Intelligent Systems and Computing

Volume 392

**Series editor**

Janusz Kacprzyk, Polish Academy of Sciences, Warsaw, Poland
e-mail: kacprzyk@ibspan.waw.pl

T0282659

## About this Series

The series "Advances in Intelligent Systems and Computing" contains publications on theory, applications, and design methods of Intelligent Systems and Intelligent Computing. Virtually all disciplines such as engineering, natural sciences, computer and information science, ICT, economics, business, e-commerce, environment, healthcare, life science are covered. The list of topics spans all the areas of modern intelligent systems and computing.

The publications within "Advances in Intelligent Systems and Computing" are primarily textbooks and proceedings of important conferences, symposia and congresses. They cover significant recent developments in the field, both of a foundational and applicable character. An important characteristic feature of the series is the short publication time and world-wide distribution. This permits a rapid and broad dissemination of research results.

## Advisory Board

More information about this series at http://www.springer.com/series/11156

Paul Chung · Andrea Soltoggio
Christian W. Dawson · Qinggang Meng
Matthew Pain
Editors

# Proceedings of the 10th International Symposium on Computer Science in Sports (ISCSS)

 Springer

*Editors*
Paul Chung
Loughborough University
Loughborough
UK

Andrea Soltoggio
Loughborough University
Loughborough
UK

Christian W. Dawson
Loughborough University
Loughborough
UK

Qinggang Meng
Loughborough University
Loughborough
UK

Matthew Pain
Loughborough University
Loughborough
UK

ISSN 2194-5357          ISSN 2194-5365   (electronic)
Advances in Intelligent Systems and Computing
ISBN 978-3-319-24558-4          ISBN 978-3-319-24560-7   (eBook)
DOI 10.1007/978-3-319-24560-7

Library of Congress Control Number: 2015950434

Springer Cham Heidelberg New York Dordrecht London

Printed on acid-free paper

Springer International Publishing AG Switzerland is part of Springer Science+Business Media
(www.springer.com)

# Preface

The 10th International Symposium of Computer Science in Sport (IACSS/ISCSS 2015), sponsored by the International Association of Computer Science in Sport and in collaboration with the International Society of Sport Psychology (ISSP), took place between September 9–11, 2015 at Loughborough, UK. Similar to previous symposia, this symposium aimed to build the links between computer science and sport, and report on results from applying computer science techniques to address a wide number of problems in sport and exercise sciences. It provided a good platform and opportunity for researchers in both computer science and sport to understand and discuss ideas and promote cross-disciplinary research.

This year the symposium covered the following topics:

- Modelling and Analysis
- Artificial Intelligence in Sport
- Virtual Reality in Sport
- Neural Cognitive Training
- IT Systems for Sport
- Sensing Technologies
- Image Processing

We received 39 submitted papers and all of them underwent strict reviews by the Program Committee. Authors of the thirty-three accepted papers were asked to revise their papers carefully according to the detailed comments so that they all meet the expected high quality of an international conference. After the conference selected papers will also be invited to be extended for inclusion in the IACSS journal.

Three keynote speakers and authors of the accepted papers presented their contributions in the above topics during the 3-day event. The arranged tour gave the participants an opportunity to see the Loughborough University campus, and facilities in the National Centre for Sport and Exercise Medicine and the Sports Technology Institute.

We thank all the participants for coming to Loughborough and hope you had enjoyed the event. We also thank the Program Committee members, the reviewers and the invited speakers for their contributions to make the event a success.

Paul Chung, General Chair
Qinggang Meng, Program Chair
Matthew Pain, Program Co-Chair

# Programme Committee

Ali Arya, Canada
Arnold Baca, Austria
Firat Batmaz, UK
Maurizio Bertollo, Italy
Bettina Bläsing, Germany
James Cochran, USA
Chris Dawson, UK
Eran Edirisinghe, UK
Hayri Ertan, Turkey
Kai Essig, Germany
Shaheen Fatima, UK
Daniel Fong, UK
Cornelia Frank, Germany
Iwan Griffiths, UK
Ben Halkon, UK
Dan Witzner Hansen, Danemark
Mark Jones, UK
Larry Katz, Canada
Rajesh Kumar, India
Martin Lames, Germany
William Land, USA
Heiko Lex, Germany
Baihua Li, UK
Keith Lyons, Australia
Andres Newball, Colombia
Jürgen Perl, Germany
Edmond Prakash, UK
Hock Soon Seah, Singapore

Thomas Schack, Germany
Didier Seyfried, France
Michael Stöckl, Austria
Martin Sykora, UK
Josef Wiemeyer, Germany
Kerstin Witte, Germany
Hui Zhang, China

## External Reviewers

Mickael Begon, Canada
Glen Blenkinsop, UK
Graham Caldwell, USA
John Challis, USA
Simon Choppin, UK
Cathy Craig, UK
Peter Dabnichki, Australia
Zac Domire, USA
Paul Downward, UK
Hayri Ertan, Turkey
Pablo Fernandez-De-Dios, UK
Sofia Fonseca, Portugal
Steph Forrester, UK
Ian Heazlewood, Australia
Ben Heller, UK
Nic James, UK
Mark King, UK
Axel Knicker, Germany
Jon Knight, UK
Daniel Link, Germany
Zhen Liu, China
Antonio Lopes, Portugal
Daniel Memmert, Germany
Toney Monnet, France
Peter O'Donoghue, UK
Kevin Oldham, UK
Leser Roland, Austria
Dietmar Saupe, Germany
Andrea Soltoggio, UK
Grant Trewartha, UK
Brian Umberger, USA

Jos Vanrenterghem, UK
Tomi Vänttinen, Finland
Sam Winter, UK
Helmut Wöllik, Austria
Jiachen Yang, China
Fred Yeadon, UK
Erika Zemkova, Slovakia

# Invited Keynote Speakers

- Prof. Arnold Baca, University of Vienna, Austria
- Dr. Michael Hiley, Loughborough University, UK
- Prof. Thomas Schack, Bielefeld University, Germany

# Contents

Part IV   Modelling and Analysis

# Part I
# Image Processing in Sport

# Non-Invasive Performance Measurement in Combat Sports

Soudeh Kasiri Behendi[1], Stuart Morgan[2], and Clinton B. Fookes[1]

[1] Queensland University of Technology, Brisbane, Australia
[2] Australian Institute of Sport, Canberra, Australia

**Abstract.** Computer vision offers a growing capacity to detect and classify actions in a large range of sports. Since combat sports are highly dynamic and physically demanding, it is difficult to measure features of performance from competition in a safe and practical way. Also, coaches frequently wish to measure the performance characteristics of other competitors. For these reasons it is desirable to be able to measure features of competitive performance without using sensors or physical devices. We present a non-invasive method for extracting pose and features of behaviour in boxing using vision cameras and time of flight sensors. We demonstrate that body parts can be reliably located, which allow punching actions to be detected. Those data can then visualised in a way that allows coaches to analysis behaviour.

## 1 Introduction

Recent advances in computer vision have enabled many examples of non-invasive measurement of performance in the sports domain including player position tracking [12, 10], and action recognition [14, 1]. Some work has also demonstrated action recognition in challenging conditions such as the aquatic environment in swimming [18, 19]. Broadly, the aim in much of the work for computer vision has been to measure features of performance without the use of invasive tracking devices or sensors. This can be described as non-invasive performance measurement. The historical alternative to non-invasive performance measurement (excluding the use of sensors or tracking devices) has been notational analysis, such that the analyst manually notates events from a competition using some predefined scheme of events and actions (See [9]). Human notational analysis, however, is notoriously vulnerable to errors such as inconsistent interpretation of event labels. Further, the manual nature of most notational analysis methods makes large-scale analyses difficult to implement. Additionally, for some dynamic and high-impact sports such as boxing, it could be dangerous for participants to wear devices of any type due to the potential risk of injury. Therefore non-invasive methods for reliable and accurate performance analysis are highly desirable.

Time of flight (ToF) sensors are a modern tool used in a range of computer vision and robotics applications where depth information is a desirable addition or replacement for conventional RGB cameras. Depth data has been widely

© Springer International Publishing Switzerland 2016
P. Chung et al. (eds.), *Proceedings of the 10th International Symposium on Computer Science in Sports (ISCSS)*, Advances in Intelligent Systems and Computing 392, DOI 10.1007/978-3-319-24560-7_1

3

used in gesture and action recognition [2–4, 17]. While computer vision has enabled many novel and exciting insights into sports performance, there are other instances where vision alone is insufficient for extract meaningful performance features, and in those instance 3D data may provide a practical solution. For instance Behendi et.al., attempted to classify punching types in boxing using overhead depth imagery [11]. In that work, punches were classified by six basic actions, *straight*, *hook* and *uppercut* (each for *rear* and *lead* hand). The direction of a boxer's forearm movement and the elbow angle were key features to determine punch types, and boxers usually throw uppercut punches from a lower initial glove position compared to hook or straight punches. Since it was not possible to differentiate between different glove positions from overhead vision alone (as illustrated in Figure 1), the main motivation for using depth data in that study was to exploit differences in the depth values of the forearm to classify uppercut punches.

(a) Lead Uppercut                          (b) Lead Hook

Fig. 1: Visual similarities between hook and uppercut punches [11].

Sports analytics research in boxing is limited, and remains a difficult problem due to the high speed of action, and occlusions in the visibility of performance features from most viewing angles. Most examples of performance analysis or activity profiling in boxing rely on slow-motion review of video footage (e.g.[5, 6]). Some efforts, such as "Box-Tag" have been made to automate scoring in boxing, which can provide additional insight to coaches about certain features of performance [7, 8]. Additionally, Morita et al [16] described a system to differentiate between punches based on gyroscopic signals providing insight on the offensive patterns of boxers.

However, despite these innovations, there are additional features of performance that are not easily extracted with existing methods. For instance, the relative position of boxers in the ring may be of significant interest to coaches, but there are no existing, non-invasive positioning methods for available for boxing. Also, the vertical movements of a boxer might be used to infer features of performance such as fatigue. Since there are no existing methods for estimating the "bouncing" of a boxer in competition, new solutions are required.

In this paper we propose a combat sports video analysis framework and demonstrate a method for extracting specific performance features in boxing using overhead depth imagery.

## 2  Methods

The general framework of the method is given in Figure 2. In this framework athletes are tracked to obtain their trajectories and analyse their movement. In the first frame, athletes are detected and trackers are assigned. Detected athletes are represented by their contour and head position. Contour tracking is used to handle partial occlusion between athletes. Finally, athletes trajectories are obtained and mapped on the ring canvas for further analysis. These stages are further described in the following sections.

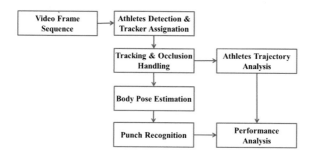

Fig. 2: Combat-sport movement analysis framework.

### 2.1  Depth Sensor

A MESA Imaging SwissRanger (SR4000) ToF sensor was used to measure activities in the boxing ring. The device was mounted approximately 6 meters above the level of the canvas ring surface. The SR4000 device generates a point cloud with a 176(h)×144(w) pixel resolution, a functional range of 10 meters, and a 69 × 55° FOV. The maximum sampling rate is 50 f.p.s. The output from the device consists of a 3×176×144 element array for calibrated distances in 3-dimensional cartesian coordinates. Viewed from above, a representation of calibrated distance values corresponding to a vision camera (mounted in tandem) is shown in Figure 3.

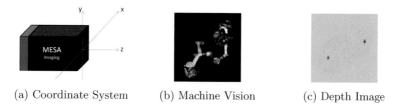

(a) Coordinate System     (b) Machine Vision     (c) Depth Image

Fig. 3: Swissranger SR-4000 Coordinate System (Courtesy Mesa Imaging AG), Machine Vision Image and Matching Depth Image.

## 2.2   Athlete Detection

This section describes the process of detecting boxers from overhead depth data.
Previous research using overhead depth data leverages the shape of the head and
shoulders for finding head candidates[13]. However, low resolution overhead data
from ToF can make detecting the those features difficult especially when their
hands are closed. A histogram of the depth data is obtained to extract the
boxing canvas depth level and depth values are translated based on the obtained
ring depth level. A precise contour of the boxer's form is obtained using the
normalised histogram of foreground contours at different depth levels (Fig. 4).
Detecting boxers head position can be challenging since boxers frequently lean

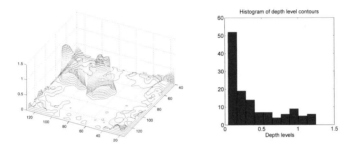

Fig. 4: Depth level contours and histogram of contour elements.

to different angles, such that the visible shape of head varies. However, detecting
the posterior location of a boxer's neck in overhead images is more reliable. A
2D chamfer distance of a boxer contour is obtained to estimate the boxer's neck
position. The properties of a boxer's contour can be "fuzzified" by assigning a
continuous probabilistic range to the boundary state, as opposed to a discrete
binary state. The neck position of the boxer is selected using the product t-norm
of fuzzified values of the candidate boxer's contour chamfer distance and depth
value [11],

$$neck = argmax_{(x,y) \in ROI}(f_z \cdot f_d),$$
$$f_z = \frac{1}{1 + e^{-a(z-c)}}, f_d = \frac{D_{ch}}{max_{(x,y) \in ROI}(D_{ch})} \tag{1}$$

where $z$ is the normalized depth value, and $D_{ch}$ is the chamfer distance computed
over the detected boxer contour.

## 2.3   Tracking and Occlusion Handling

Once candidate boxers have been detected in first frame, they can then be
tracked over consecutive frames to obtain a continuous movement trajectory.

Boxers are non-rigid objects and occlude each other frequently (Fig. 5). The boxers' head and shoulders are relatively stable features and provide continuity of position over successive frames such that contour tracking can be used to obtain boxers trajectories. Contour tracking handles topological changes, such as merging and splitting of object regions. When occlusion occurs, the contours of the athletes are merged. At the end of the occlusion, the group contour is split and each athlete is tracked individually. The main problem in an occlusion situation is identifying each boxer and determining their positions after the occlusion. Although boxers occlude each other and their contours are merged, it is usually partial occlusion from overhead view. Regional maxima of the $f_z \cdot f_d$ for the merged contours are obtained and neck positions are estimated, which is illustrated in Fig. 5. Detected heads in occluded contour are shown by red and pink points in Fig. 5(e).

Robust position tracklets can then be derived using the calibrated x-y position coordinate system provided by the raw data files. Tracking data is then retained in the form of frame-based rows, each consisting of X, Y, Z cartesian coordinates where the origin is at the canvas level in the approximate centre of the ring.

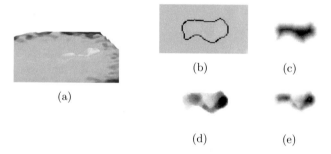

(b)          (c)

(a)

(d)          (e)

Fig. 5: Intermediate results of boxers detection: (a) the depth 3d mesh, (b) contour of the merged boxers, (c) $f_d$, (d) $f_z$, and (e) $f_z \cdot f_d$.

## 2.4 Athletes Movement Analysis

Performance Analysts for combat sports are frequently interested in physical proximity of two boxers, and the extent to which they each move in and out of an effective striking range. Using the position estimates extracted using the methods described above the momentary distance between the boxers can be derived as a 3-D Euclidean distance using:

$$dt(p,q) = \sqrt{\sum_{i=1}^{3}(p_i - q_i)^2} \qquad (2)$$

Local point values can be visualised for coaching purposes using a bespoke interactive visualisation tool developed using OpenGL at the Australian Institute of Sport [15]. Exemplar results are shown in Figure 6.

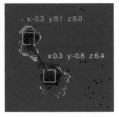

Fig. 6: Tracking and Inter-boxer distance estimates derived from position tracking.

## 3   Results

We evaluated our method using a sequence of depth arrays taken at the Australian Institute of Sport from boxing sparring. A time series of the inter-boxer distance can be calculated for greater understanding of the fluctuations in proximity between boxers as a function of various actions and behaviours (Figure 7a). In this instance the raw distance estimates are smoothed using the Tukey's (Running Median) Smoothing function: smooth {stats}, in R (version 3.2.0). Similarly, the vertical oscillations of two boxers in sparring may be related to evidence of physical fatigue. As such performance analysts are interested in monitoring the amount of "bouncing" that occurs over time in a bout. These data can be simply extracted as time series data from the calibrated z-axis, and an exemplar is show in Figure 7b. Discrete estimates of the degree of vertical oscillations could be further derived using measures of dispersion over a sample, or to analyse the data in the frequency domain.

## 4   Conclusions

Computer vision is becoming increasingly important in sports analytics as a non-invasive method for extracting the occurrence of actions in competition, and for understanding the features of sports performance without impacting on the performance environment with physical motion sensors. Boxing and combat sports represent a particularly challenging domain for action recognition with vision, and demonstrate a method for extracting features of performance using ToF sensors. Our results demonstrate that it is possible to track multiple boxers in a sparring contest, and to extract additional features including punch types, ring position, vertical movement, and the inter-boxer distance. Future work will aim to integrate previous punch classification work with athlete positioning to demonstrate a unified performance analysis system.

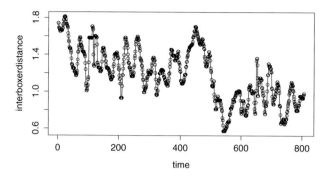

(a) Time series analysis of exemplar inter-boxer distances in sparring.

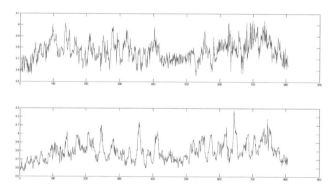

(b) Time series analysis of exemplar vertical oscillations for two boxers.

Fig. 7: Performance feature extractions from boxer head/neck tracking.

## 5 Acknowledgments

The authors gratefully acknowledge the support from the Australian Institute of Sport Combat Centre, and in particular from Emily Dunn, Michael Maloney, Clare Humberstone, and David T Martin.

## References

1. Agarwal, A., Triggs, B.: Recovering 3D human pose from monocular images. Pattern Analysis and Machine Intelligence, IEEE Transactions on 28(1), 44–58 (Jan 2006)
2. Aggarwal, J., Xia, L.: Human activity recognition from 3d data: A review. Pattern Recognition Letters (2014)

3. Baak, A., MÃŒller, M., et al.: A data-driven approach for real-time full body pose reconstruction from a depth camera. In: Consumer Depth Cameras for Computer Vision, pp. 71–98. Springer (2013)
4. Chen, L., Wei, H., Ferryman, J.: A survey of human motion analysis using depth imagery. Pattern Recognition Letters 34(15), 1995 – 2006 (2013), smart Approaches for Human Action Recognition
5. Davis, P., Benson, P., Waldock, R., Connorton, A.: Performance analysis of elite female amateur boxers and comparison to their male counterparts. International journal of sports physiology and performance (2015)
6. Davis, P., Benson, P.R., Pitty, J.D., Connorton, A.J., Waldock, R.: The activity profile of elite male amateur boxing. International journal of sports physiology and performance (10), 53–57 (2015)
7. Hahn, A., Helmer, R., Kelly, T., Partridge, K., Krajewski, A., Blanchonette, I., Barker, J., Bruch, H., Brydon, M., Hooke, N., et al.: Development of an automated scoring system for amateur boxing. Procedia Engineering 2(2), 3095–3101 (2010)
8. Helmer, R., Hahn, A., Staynes, L., Denning, R., Krajewski, A., Blanchonette, I.: Design and development of interactive textiles for impact detection and use with an automated boxing scoring system. Procedia Engineering 2(2), 3065–3070 (2010)
9. Hughes, M., Franks, I.M.: Notational analysis of sport: Systems for better coaching and performance in sport. Psychology Press (2004)
10. Kasiri-Bidhendi, S., Safabakhsh, R.: Effective tracking of the players and ball in indoor soccer games in the presence of occlusion. In: 14th International CSI Computer Conference. pp. 524–529. IEEE (2009)
11. Kasiri-Bidhendi, S., Fookes, C., Morgan, S., Martin, D.T.: Combat sports analytics: Boxing punch classification using overhead depth imagery. In: Image Processing (ICIP), 2015 IEEE International Conference on (2015)
12. Liu, J., Carr, P., Collins, R.T., Liu, Y.: Tracking sports players with context-conditioned motion models. In: Computer Vision and Pattern Recognition (CVPR), 2013 IEEE Conference on. pp. 1830–1837. IEEE (2013)
13. Migniot, C., Ababsa, F.: 3d human tracking in a top view using depth information recorded by the xtion pro-live camera. In: Advances in Visual Computing, pp. 603–612. Springer (2013)
14. Moeslund, T.B., Hilton, A., Krüger, V.: A survey of advances in vision-based human motion capture and analysis. Computer vision and image understanding 104(2), 90–126 (2006)
15. Morgan, S.: 3dviewkit, openGL software application; Australian Institute of Sport.
16. Morita, M., Watanabe, K., et al.: Boxing punch analysis using 3D gyro sensor. In: SICE Annual Conference (SICE), 2011 Proceedings of. pp. 1125–1127 (Sept 2011)
17. Munaro, M., Basso, A., Fossati, A., Van Gool, L., Menegatti, E.: 3d reconstruction of freely moving persons for re-identification with a depth sensor. In: Robotics and Automation (ICRA), 2014 IEEE International Conference on. pp. 4512–4519 (May 2014)
18. Sha, L., Lucey, P., Morgan, S., Pease, D., Sridharan, S.: Swimmer localization from a moving camera. In: Digital Image Computing: Techniques and Applications (DICTA), 2013 International Conference on. pp. 1–8. IEEE (2013)
19. Sha, L., Lucey, P., Sridharan, S., Morgan, S., Pease, D.: Understanding and analyzing a large collection of archived swimming videos. In: Applications of Computer Vision (WACV), 2014 IEEE Winter Conference on. pp. 674–681. IEEE (2014)

# Comparison between Marker-less Kinect-based and Conventional 2D Motion Analysis System on Vertical Jump Kinematic Properties Measured from Sagittal View

Shariman Ismadi Ismail*, Effirah Osman, Norasrudin Sulaiman, Rahmat Adnan

Faculty of Sports Science & Recreation, Universiti Teknologi MARA, 40450 Shah Alam, Selangor, Malaysia

shariman_ismadi@salam.uitm.edu.my / shariman_2000@yahoo.com

**Abstract.** Marker-less motion analysis system is the future for sports motion study. This is because it can potentially be applied in real time competitive matches because no marking system is required. The purpose of this study is to observe the suitability and practicality of one of the basic marker-less motion analysis system applications on human movement from sagittal view plane. In this study, the movement of upper and lower extremities of the human body during a vertical jump act was chosen as the movement to be observed. One skilled volleyball player was recruited to perform multiple trials of the vertical jump (n=90). All trials were recorded by one depth camera and one Full HD video camera. The kinematics of shoulder joint was chosen to represent the upper body extremity movement while knee joint was chosen as the representative of the lower body extremity movement during the vertical jump's initial position to take-off position (IP-TP) and take-off position to highest position (TP-HP). Results collected from depth camera-based marker less motion analysis system were then compared with results obtained from a conventional video-based 2-D motion analysis system. Results indicated that there were significant differences between the two analysis methods in measuring the kinematic properties in both lower (knee joint) and upper (shoulder joint) extremity body movements ($p < .05$). It was also found that a lower correlation between these two analysis methods was more obvious for the knee joint movement [38.61% matched, $r = 0.12$ (IP-TP) and $r = 0.01$ (TP-HP)] compared to the shoulder joint movement [61.40% matched, $r = 0.10$ (IP-TP) and $r = 0.11$ (TP-HP)].

**Keywords:** Motion analysis, marker less, depth camera, vertical jump

## 1 Introduction

The capture technique of human movement is one of the crucial parts recently used by biomechanics to study the musculoskeletal movement and is also being used by physiologists to diagnose an injury problem. According to Krosshaug et al (2007), analysis of human motion is very useful in establishing injury risks through joint position

© Springer International Publishing Switzerland 2016
P. Chung et al. (eds.), *Proceedings of the 10th International Symposium on Computer Science in Sports (ISCSS)*, Advances in Intelligent Systems and Computing 392, DOI 10.1007/978-3-319-24560-7_2

measurement and orientation of body segments, as well as analyzing the technique in sports (Lees, 2002). Thus, it is important to obtain the robustness and accuracy of the results in order to detect every single motion involved in particular human movements.

There are several approaches recommended for use as simple setup tools in order to obtain stable, accurate, and real-time motion capturing performances. The marker-based system tool has been proven to be suitable for in-vitro (laboratory based) studies where the subject has to wear an obstructive device, a marker which is more complicated, hard to maintain, and even quite expensive. Although the application demands to use these tools have increased during a real-time competitive sporting event, but it is difficult for athletes to do their normal routines with the marker placed on their body (Wheat, Fleming, Burton, Penders, Choppin, & Heller, 2012; Zhang, Sturm, Cremers, & Lee, 2012).

The marker-less based system tools have come out with attractive solutions to solve problems associated with marker-based system tools. Microsoft launched the low cost marker-less camera-based Kinetics which originally was used for Xbox 360 gaming, with the capability for tracking the users' body segment positions and 3D orientations in real situations. These cameras require minimal calibration by standing in a specific position only for a few seconds with no marker required on the body. However, this tool also has their own limitations resulting in low accuracy and less supported on motions with high speed (Choppin & Wheat, 2012). With a lower price compared to other depth camera, these cameras are only capable of capturing 30 frames per second. It means that these cameras have the capability only for capturing certain basic motions or movements like walking or jumping, rather than fast movements (Corazza et al., 2006; Zhang et al., 2012).

Therefore, this study was designed to observe the suitability and practicality of depth camera applications in vertical jump focusing on upper and lower extremity body movements when located at the sagittal plane with respect to the movement.

## 2    Method

### 2.1    Subject

One skilled amateur volleyball player (age 24 years, height 178 cm, weight 75 kg with 10 years of competitive volleyball playing experience) was recruited to participate in this study. Consent from the subject and approval from the research ethics committee from the research organization was obtained before the study was conducted.

## 2.2    Instrumentation

One depth camera (Microsoft's Kinect) and one Full HD video camera (Sony-60 FPS) were utilized in this study. The depth camera has the capability of depth data capture at 30FPS with a resolution of 640 x 480 pixels. It is capable of tracking various types of joint angles. Depth Biomechanics by Sheffield Hallam University (Depth Biomechanics, 2015) was the software utilized in this study to process the data captured by the depth camera. KINOVEA software (v. 0.8.15) (Kinovea, 2015) was used to analyze the video. Two units of reflective markers (d=14mm) were located at the right side of the subject's shoulder and knee joint. Subject was asked to perform warm up and stretching exercises for 5 to 10 minutes prior to performing the jump. All cameras were set at the sagittal view of the subject's body, as shown in Fig. 1. Typical calibration of the depth camera (front view calibration) was performed before the recording took place.

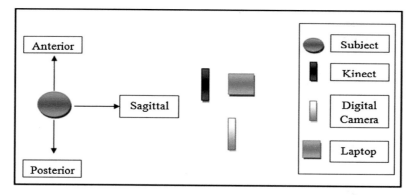

**Fig. 1.** Instrumentation setup (Top View)

## 2.3    Data collection and processing

In this study, the shoulder joint was chosen as a representative of upper body extremity, while the knee joint represents the lower body extremity. From these two major parts, each part consists of two different types of phases to be analyzed, which include initial position to take-off phase (IP-TP), and take-off position to the highest phase (TP-HP), as shown in Fig. 2.

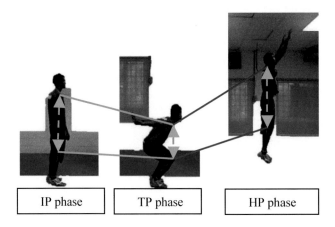

| IP phase | TP phase | HP phase |

**Fig. 2.** Vertical Jump phases used to measure displacement

According to Adams and Beam (2008), proper jumping maneuver for the vertical jump was 3 trials with about 20 – 30 s of recovery between trials. Each of these 3 trials is considered as 1 set of jumps. Throughout this study, the subject performs 35 sets of jumps. Between each set, the subject will rest between 1-3 minutes. From all trials recorded, 90 trials were selected for further analysis.

Synchronization of the frame rate between the video camera-based and depth camera-based data were required prior to data analysis. In order to synchronize it, the time frame and joint displacement obtained from the video analysis by Kinovea were converted to coordinate system, to make it similar to the output obtained from Depth Biomechanics. This study tends to evaluate the two methods of analysis at its most optimum setting. Therefore the video camera was set at full high definition resolution (1920 x1080) with 60 FPS. Since the data from the depth camera was 30 FPS (resolution at 640 x 480), therefore the time-frame rate need to be adjusted for both sources to be in-sync before the results were analyzed. Finite forward difference method was utilized in order to calculate the joint displacement based on the coordinates obtained from both depth and video cameras. Independent t-test was used to compare data from the two methods of analysis. Lastly, correlation analysis was performed to observe how strong the results between the two methods of analysis were similar.

## 3    Results and Discussion

The result of the study showed that there were significant differences between the two methods analyzed in all jumping phases (Table 1). It was also found that a lower correlation between these two analysis methods were more obvious for the knee joint movement [r= 0.12 (IP-TP) and r=0.01 (TP-HP)] compared to the shoulder joint movement [r=0.10 (IP-TP) and r=0.11 (TP-HP)]. This result can be explained based on the differences in speed in each of the phases. Although it was not shown here, but the speed of TP-HP was faster than the movements during the IP-TP phases. It is also worthwhile mentioning that based on point-to-point data comparison in each trial, the data for knee joint displacement (lower body movement) only achieved 38.61% matching results and the data from the shoulder joint displacement (upper body movement) achieved 61.40% matching results when compared between the two types of analysis methods.

Microsoft Kinect depth camera has its own limitation resulting in low accuracy and less support on motions with high speed (Choppin & Wheat, 2012). With a lower price than the other depth cameras, these cameras have capabilities only for capturing certain basic motions or movements like walking or jumping, rather than fast movements (Corazza et al., 2006; Zhang et al., 2012). This study shows that the accuracy level reduces when the depth camera measures a higher speed motion.

These types of depth camera with capturing capabilities of 30 frames per second at 640 x 480 resolutions are more accurate when used at the frontal plane. In gaming situations, the upper extremity is used more frequently rather than the lower extremity. Also, the camera is typically located in the frontal view, and not from the sagittal view. The calibration procedure was also only based on frontal view calibration and not the sagittal view.

**Table 1.** Comparison of Displacement Measurement between 2D Video-based and Marker-less Motion Analysis.

| Segment | Method | N (No. of trial) | Mean, cm (SD) | t (t-value) | *Correlation* (r) |
|---|---|---|---|---|---|
| Upper body IP-TP Displacement (Shoulder joint) | 2D-Video | 90 | 54.85 (3.21) | 6.3* | 0.10 |
| | Marker-less | 90 | 141.88 (131.27) | | |
| Upper body TP-HP Displacement (Shoulder joint) | 2D-Video | 90 | 118.78 (3.87) | 7.2* | 0.11 |
| | Marker-less | 90 | 252.38 (175.42) | | |

| Lower body IP-TP Displacement (Knee joint) | 2D-Video | 90 | 24.00 (5.65) | 3.5* | 0.12 |
|---|---|---|---|---|---|
| | Marker-less | 90 | 62.42 (104.80) | | |
| Lower body TP-HP Displacement (Knee joint) | 2D-Video | 90 | 74.39 (16.48) | 6.4* | 0.01 |
| | Marker-less | 90 | 196.23 (179.79) | | |

*Mean values between the two method are significantly different (p < 0.05)

If a depth camera is to be utilized from sagittal view in a motion analysis involving rapid movement, a higher specification of the camera in terms of the frame rate should be considered. There must also be a calibration procedure involving sagittal view. However, further studies are still needed to suggest that a higher frame rate and additional calibration steps could improve data captured from other viewing angles or planes, beside the frontal view. Also, the resolution of the camera could also be a factor that influenced accuracy. A depth camera with a higher resolution, higher than 640x480, might provide an improvement to the marker-less motion analysis method.

## 4   Conclusion

From this study, it can be concluded that video-based analysis and depth camera-based analysis have significant differences on the upper extremity and lower extremity towards the vertical jump motion, when captured from the sagittal view. In terms of accuracy between the upper body and the lower body based analysis by the Microsoft Kinect depth camera, the upper body was slightly more accurate than the lower body. A depth camera with 30 FPS was less suitable when capturing fast movements, especially from the sagittal view plane. It would be interesting to note whether these types of tools can come out with high level of accuracy towards other movements such as kicking or throwing if the frame rate was to be improved. In order to get higher accuracy, it is recommended that 2 or more Microsoft Kinect depth cameras be used to detect the full capturing angles; frontal plane and sagittal planes. By using more than one depth camera, the validity of the 3D data capturing ability can be observed from different viewpoints.

**Acknowledgment**
    This work was supported by the Malaysian Ministry of Education's Sports Grant (Geran Sukan KPT) [100-RMI/GOV 16/6/2 (3/2014) ]

# References

1. Krosshaug, T., Slauterbeck, J. R., Engebretsen, L., & Bahr, R. (2007). Biomechanical analysis of anterior cruciate ligament injury mechanisms: Three-dimensional motion reconstruction from video sequences. *Scandinavian Journal of Medicine & Science in Sports, 17*(5), 508-519. doi: 10.1111/j.1600-0838.2006.00558.x
2. Lees, A. (2002). Technique analysis in sports: A critical review. *Journal of Sports Sciences, 20*(10), 813-828. doi: 10.1080/026404102320675657
3. Wheat, J., Fleming, R., Burton, M., Penders, J., Choppin, S., & Heller, B. (2012). Establishing the accuracy and feasibility of microsoft kinect in various multidisciplinary contexts. *EFL Kinect: Project Report.*
4. Zhang, L., Sturm, J., Cremers, D., & Lee, D. (2012). Real-time human motion tracking using multiple depth cameras Retrieved from https://www.zotero.org/groups/3d_mapping_with_kinect/items/itemKey/AMEVV4XS
5. Choppin, S., & Wheat, J. (2012). *Marker-less tracking of human movement using microsoft kinect.* Paper presented at the 30th Anual Conference of Biomechanics in Sports.
6. Corazza, S., Mundermann, L., Chaudhari, A. M., Demattio, T., Cobelli, C., & Andriacchi, T. P. (2006). A markerless motion capture system to study musculoskeletal biomechanics: Visual hull and simulated annealing approach. *Annals of Biomedical Engineering,, 34*(6), 1019–1029. doi: 10.1007/s10439-006-9122-8
7. Depth Biomechanics manual.2015. http://www.depthbiomechanics.co.uk/?s=manual (Acessed 2015-07-07)
8. KINOVEA manual.2015. http://www.kinovea.org/help/en/index.htm (Accessed 2015-07-07)
9. Adams, G. M., & Beam, W. C. (2008). Maximal oxygen consumption. In C. Johnson (Ed.), *Exercise physiology laboratory manual* (5th ed., pp. 149-168). New York: McGraw-Hill.

# Test of ball speed in table tennis based on monocular camera

Hui Zhang[1], Ling-hui Kong[2], Ye-peng Guan[3] and Jin-ju Hu[4]

[1] Sport Science Department, College of Education, Zhejiang University, Hangzhou 310028, China
[2] China Table Tennis College, Shanghai University of Sport, Shanghai 200438, China
[3] School of Communication & Information Engineering, Shanghai University, Shanghai 200444, China
[4] School of Sport Journalism and Foreign Studies, Shanghai University of Sport, Shanghai 200438, China

**Abstract:** This paper has designed and developed a platform for testing table tennis ball speed, which are used in the tests of 5 top Chinese female table tennis players in three of their practices. The results show that the ball speeds of LI and DING are faster than those of LIU, ZHU and CHEN. LI's ball speed over the net is the fastest in all exercises, especially in forehand loop-drive against backspin and forehand moving loop-drive. DING's ball speed over the net is the second-fastest in different practices (except for forehand loop-drive in the Two to One practice). LIU, ZHU and CHEN's ball speeds are slower, among which, the relatively slower ball speeds over the net were LIU's backhand loop-drive in the Two to One practice, ZHU's forehand loop-drive in the Two to One exercise and in the forehand loop-drive against backspin, and CHEN's forehand moving loop-drive.

**Key words:** Table tennis, loop-drive, ball speed over the net, ball speed rebounded from the table

## 1 Introduction

Speed and spin are the two most important properties in table tennis sport. WU and QIN et al (1988) have conducted tests on table tennis spins using their self-developed testing instrument. SUN and YU et al (2008), through solving the math models of table tennis sport, have analyzed the moving path of loop-drives and their general rules of movement in different circumstances. FANG (2003), JIANG and LI et al (2008), FANF and ZHANG et al (2011) and YANG and YUAN et al (2014), through table tennis simulation, have conducted researches on table tennis collision process, flying path and bouncing features. However, due to technical reasons, the testing methods on spins and speeds accomplished in the lab, or the researches on the flying path and bouncing speed based on simulation cannot be widely used in real life practice. Hence, the present research has designed and developed a platform for testing the

© Springer International Publishing Switzerland 2016
P. Chung et al. (eds.), *Proceedings of the 10th International Symposium on Computer Science in Sports (ISCSS)*, Advances in Intelligent Systems and Computing 392, DOI 10.1007/978-3-319-24560-7_3

ball speed of table tennis players using the monocular camera, which has already been put into practice in female elite table tennis players.

## 2    Methods

### 2.1    Participants

The five top players from Chinese national women team (Xiao-xia LI, Ning DING, Shi-wen LIU, Yu-ling ZHU and Meng CHEN, all world champions) participated in this test during the assembled training session.

### 2.2    Testing method of ball speed

**Table-Camera mapping relation**

The location of each point in the flat surface of the camera image is related to the its geometric position of the correspondent object in the three dimensional space. In other words, the spacial location of a three-dimensional object is closely correspondent to its planimetric position in the two-dimensional image. The correspondent relation is determined by the geometric model of the camera. Therefore, based on the differences between the physical features of the three dimensional table and its background image, the visual features can be obtained of the edges of the table and the net, and the mapping model can be constructed of the three dimensional table and the planimetric camera imaging.

**Determining the parameters of camera imaging**

Based on the constructed mapping model of the three dimensional table and the planimetric camera imaging, the image coordinate system, camera coordinate system and the world coordinate system concerned with the parameters involved in camera imaging have the following relationship:

$$s \begin{bmatrix} u \\ v \\ 1 \end{bmatrix} = \begin{bmatrix} m_1 & m_2 & m_3 & m_4 \\ m_5 & m_6 & m_7 & m_8 \\ m_9 & m_{10} & m_{11} & m_{12} \end{bmatrix} \cdot \begin{bmatrix} X \\ Y \\ Z \\ 1 \end{bmatrix} \tag{1}$$

Among them, $s$ is the arbitrary number except 0; $u$, $v$ are the respective pixel coordinate of the three dimensional points mapped onto the camera plane; $m_i$ ($i$=1, 2,..., 12) are the camera projection matrix; $X$, $Y$, $Z$ are the respective world coordinate of the three dimensional space points.

Using four or the above mentioned three-dimensional points ($X$, $Y$, $Z$) on the table tennis platform and their correspondent pixels in the camera plane imaging ($u$, $v$) and based on singular value equation, determine the geometrical parameter during the camera plane imaging $m_i$ ($i$=1, 2,..., 12).

**Extraction of table tennis visual features**

Due to the complicated and changeable background and circumstances in table tennis training and competition venues, and also because of the uncontrollable and unpredictable outer environmental factors (lights, people in and out), the venue background needs to be self adaptively updated. Since players in the foreground have higher range of movement features, and any of the observed moving targets in the foreground can be reflected in the changes of sequence of scene images, therefore, the pixel range of the adjacent foreground targets can be obtained. If the differences between the obtained adjacent objects exceed certain range, it shows that the background has changed and needs to be self updated.

Based on the self-adaptively updated background, using methods of multi-scale wavelet and particle dynamics, the moving video objects are segmented and the table tennis foreground objects are extracted from the video pictures. Then according to the color ratio invariance property of table tennis foreground object, the ghost shadows extracted from balls' flying path are suppressed, which has overcome many defects in the current segmentation methods of foreground moving objects in videos, such as defects of manual correction or human judgment, priori hypothesis, as well as the sensitivity towards dynamic scenes and noise interventions.

**Table tennis ball three dimensional coordinates**

Based on the extracted table tennis ball visual features, the visual pixel points are extracted about the table tennis balls in the up, down, left and right edges. And according to the geometrical parameters $m_i$ ($i=1,2,...12$) established during the camera plane imaging and the geometrical invariance nature in the camera plane imaging, eight linear equations are established with formula (1). With the least square methods, the three dimensional coordinates ($X, Y, Z$) are to be calculated of the table tennis ball central points.

**Establishment of table tennis ball speed/accelerated speed**

And the ball's three dimensional coordinates established in different video times, the flying distance of the ball is calculated, and then its flying speed as well as its accelerated speed, according to the camera video frame rate established by camera encoding and decoding.

## 2.3 Experiment

The experiment was conducted at the second-floor training hall in the Chengdu Table Tennis Athletic School (National Table Tennis Training Base) during March 10 and April 20, 2014. The tests were done in the second, seventh and tenth units of the assembled training session. The whole training process of the five players on the three training items were recorded and used for analysis with the specially developed table tennis software. And the real time speed was captured of the ball over the net when the players stroke.

**Test contents**

In order to reflect the real life striking speed of the players, the coach group of the national women table tennis teams had serious discussions and finally decided on the following three training items for testing:

1. Two to One method: It refers to the strategy that two practice team members train together with one tested player. The practice team member strikes the ball to the full court of the tested player and the tested player strikes back with forehand or backhand loop-drive.
2. Forehand loop-drive against backspin: It refers to the practice that one coach serves backspin ball (multi-ball practice), and the tested player strikes back with forehand loop-drive in full court.
3. Forehand moving loop-drive: It refers to the strategy that the one practice team member strikes the ball to the forehand, mid-route and backhand of the tested player, and the tested player strikes back with forehand loop-drives.

**Test method and data processing.**
The complete training period of the above five players were recorded and analyzed using the special table tennis software, and the instant speed is obtained when the ball crossed the net. The ball speed analysis is done on the experimental platform which catches and stores the ball speeds when crossing the net and bouncing after crossing the net. The platform is developed by the intellectual information sensory lab of Shanghai University under the environment of VC++2010, which can be operated in Window 7 system with 32 units (Figure 1).

Fig. 1. The platform of table tennis ball speed testing

# 3    Results

## 3.1    General features of striking speed of top female players

The general features of the ball speed after striking of top female players are shown in table 1. Among all the tests on each training item, the net crossing speed and bouncing speed after crossing the net are the quickest with the forehand loop-drive against backspin, 12.245 m/s and 10.966 m/s respectively. The forehand moving loop-drive comes the second, 11.589 m/s and 10.316 m/s each. And the third comes with the forehand loop-drive in the Two to One practice, with the speed of 9.049 m/s and 7.968 m/s respectively. In the forehand loop-drive against backspin practice, the players have enough time adjusting their positions, so they can take the time driving and pulling powerfully, and hence resulting in the quickest net crossing speed and bouncing speed after crossing the net.

Table 1.  Data on General features of ball speed of elite female players' striking

|  | N | Ball speed over the net (m/s) | | | |
|---|---|---|---|---|---|
|  |  | Mean | Sd. | Min. | Max. |
| Two to One method (forehand loop-drive) | 829 | 10.675 | 2.134 | 3.755 | 16.705 |
| Two to One method (backhand loop-drive) | 1147 | 9.049 | 1.470 | 3.552 | 14.498 |
| Forehand loop-drive against backspin | 822 | 12.245 | 1.763 | 5.938 | 18.118 |
| Forehand moving loop-drive: | 2010 | 11.589 | 1.598 | 4.831 | 16.559 |
|  | N | Ball speed rebounded from the table (m/s) | | | |
|  |  | Mean | Sd. | Min. | Max. |
| Two to One method (forehand loop-drive) | 829 | 9.473 | 2.051 | 2.534 | 14.825 |
| Two to One method (backhand loop-drive) | 1147 | 7.968 | 1.454 | 3.109 | 13.512 |
| Forehand loop-drive against backspin | 822 | 10.966 | 1.759 | 5.773 | 17.650 |
| Forehand moving loop-drive: | 2010 | 10.316 | 1.660 | 3.549 | 14.957 |

In the present test, the largest net crossing speed and after-net bouncing speed are 18.118 m/s and 17.650 m/s respectively (forehand loop-drive against backspin), and the smallest are 3.552 m/s (Two to one method: backhand loop-drive) and 2.534 m/s (Two to one method: forehand loop-drive). This big difference lies in that, in the first drive in each round, the players often use little effort because they have to adjust their positions, sometimes in the process of striking, they also need to adjust themselves according to the placement, speed, spin of the coming ball. This is why there is such great difference.

Besides, because the forehand striking in table tennis belongs to the open skill and the backhand striking the closed skill, the backswing and power with the forehand would be better than the backhand. Therefore, both the net-crossing speed and the after-net bouncing speed of the forehand loop-drives are faster than those of backhand loop-drives. On the other hand, the stability of the closed technique is better than the open technique, so the standard deviation of net-crossing speed and the after-net bouncing speed of the backhand loop-drives is smaller than that of the forehand loop-drive.

## 3.2    Comparison of ball speed of forehand loop-drive in the Two to One practice

Table 2.  Data on ball speed of forehand loop-drive in the Two to One practice

|  | N | Ball speed over the net (m/s) | | | |
|---|---|---|---|---|---|
|  |  | Mean | Sd. | Min. | Max. |
| LI | 216 | 10.911 | 2.064 | 4.063 | 15.152 |
| DING | 136 | 10.534 | 2.023 | 5.260 | 15.143 |
| LIU | 165 | 10.728 | 1.324 | 6.630 | 13.477 |
| ZHU | 186 | 10.349 | 2.975 | 3.755 | 16.705 |
| CHEN | 126 | 10.834 | 1.646 | 6.613 | 14.571 |
|  | N | Ball speed rebounded from the table (m/s) | | | |
|  |  | Mean | Sd. | Min. | Max. |
| LI | 216 | 9.567 | 2.031 | 3.307 | 14.825 |
| DING | 136 | 9.598 | 1.932 | 4.072 | 14.141 |
| LIU | 165 | 9.475 | 1.468 | 6.273 | 12.809 |
| ZHU | 186 | 9.170 | 2.738 | 2.534 | 13.992 |
| CHEN | 126 | 9.623 | 1.623 | 6.257 | 13.282 |

The net-crossing speed with forehand loop-drives in the Two to One practice ranks as this: LI > CHEN > LIU > DING > ZHU, while the ranking of the after-net bouncing speed goes as: CHEN > DING > LI > LIU > ZHU (Table 2). The inconsistency of the rankings between the net-crossing speed and the after-net bouncing speed may be resulted from the spinning of the loops. The present research didn't monitor over the ball spinning speed of the players, therefore, the reason for this phenomenon cannot be fully explained. But the statistical test proves that the above differences have no significance ($F=2.106$, $P>0.05$).

### 3.3 Comparison of ball speed of backhand loop-drive in the Two to One practice

Table 3. Data on ball speed of backhand loop-drive in the Two to One practice

|  | N | Ball speed over the net (m/s) | | | |
|---|---|---|---|---|---|
|  |  | Mean | Sd. | Min. | Max. |
| LI | 292 | 9.391 | 1.667 | 3.552 | 13.799 |
| DING | 195 | 9.310 | 1.217 | 5.544 | 13.104 |
| LIU | 282 | 8.607 | 1.257 | 4.611 | 11.691 |
| ZHU | 190 | 9.043 | 1.878 | 3.754 | 14.498 |
| CHEN | 188 | 8.913 | 0.906 | 5.451 | 11.033 |
|  | N | Ball speed rebounded from the table (m/s) | | | |
|  |  | Mean | Sd. | Min. | Max. |
| LI | 292 | 8.098 | 1.701 | 3.178 | 12.526 |
| DING | 195 | 8.269 | 1.281 | 4.052 | 13.033 |
| LIU | 282 | 7.602 | 1.154 | 3.934 | 11.093 |
| ZHU | 190 | 7.977 | 1.792 | 3.109 | 13.512 |
| CHEN | 188 | 7.996 | 1.091 | 4.665 | 10.655 |

### 3.4 Comparison of ball speed in forehand loop-drive against backspin practice

The net-crossing speed and the after-net bouncing speed with forehand loop-drive against backspin are showed in table 4. The net-crossing speed of CHEN, LIU and ZHU is obviously slower than that of DING and LI ($F=28.577$, $P<0.01$), showing very significant differences. Their after-net bouncing speed is also obviously slower than that of DING and LI. Besides, DING's ball speed is clearly slower than that of LI, also showing very significant differences ($F=31.561$, $P<0.01$).

Table 4. Data on ball speed in forehand loop-drive against backspin practice

|  | N | Ball speed over the net (m/s) | | | |
|---|---|---|---|---|---|
|  |  | Mean | Sd. | Min. | Max. |
| LI | 152 | 13.202 | 1.820 | 5.938 | 16.651 |
| DING | 154 | 12.832 | 1.329 | 9.009 | 15.701 |
| LIU | 137 | 11.695 | 1.500 | 6.969 | 15.004 |
| ZHU | 185 | 11.645 | 1.925 | 6.440 | 17.416 |
| CHEN | 194 | 11.987 | 1.578 | 8.150 | 18.118 |
|  | N | Ball speed rebounded from the table (m/s) | | | |
|  |  | Mean | Sd. | Min. | Max. |
| LI | 152 | 11.988 | 1.640 | 5.773 | 15.589 |
| DING | 154 | 11.524 | 1.609 | 6.869 | 14.779 |

| LIU  | 137 | 10.482 | 1.381 | 6.495 | 14.084 |
| ZHU  | 185 | 10.268 | 1.710 | 6.135 | 14557 |
| CHEN | 194 | 10.729 | 1.763 | 6.024 | 17.650 |

### 3.5 Comparison of ball speeds in forehand moving loop-drive practice

Table 5 shows the net-crossing speed and the after-net bouncing speed of LI are the fastest in the forehand moving loop-drive practice. Among them, the net-crossing speed and the after-net bouncing speed of CHEN, LIU and ZHU are obviously slower than those of DING and LI, and the net-crossing speed and the after-net bouncing speed of DING are also clearly slower than that of LI ($F=26.626$, $P<0.01$; $F=25.481$, $P<0.01$) showing very significant differences.

Table 5. Data on ball speed in forehand moving loop-drive practice

|      | N | Ball speed over the net (m/s) | | | |
|      |   | Mean | Sd. | Min. | Max. |
|------|---|------|-----|------|------|
| LI   | 318 | 12.262 | 1.859 | 5.167 | 16.559 |
| DING | 396 | 11.825 | 1.784 | 5.365 | 15.711 |
| LIU  | 512 | 11.394 | 1.341 | 4.831 | 14.312 |
| ZHU  | 449 | 11.456 | 1.399 | 6.733 | 14.547 |
| CHEN | 335 | 11.148 | 1.469 | 6.362 | 14.544 |
|      | N | Ball speed rebounded from the table (m/s) | | | |
|      |   | Mean | Sd. | Min. | Max. |
| LI   | 318 | 11.020 | 2.067 | 4.073 | 14.957 |
| DING | 396 | 10.469 | 1.652 | 5.186 | 13.994 |
| LIU  | 512 | 10.188 | 1.456 | 3.549 | 12.550 |
| ZHU  | 449 | 10.166 | 1.427 | 5.600 | 13.084 |
| CHEN | 335 | 9.830 | 1.590 | 4.506 | 13.297 |

## 4    Conclusion

The platform designed and developed in this paper for measuring table tennis ball speeds can be conveniently used in elite players training. In the above five top players, the ball speeds of LI and DING are faster than those of LIU and ZHU and CHEN. LI ranks No. 1 in ball speed in all training practices, especially in her forehand loop-drive against backspin and moving loop-drive practice. Ding ranks the second in the net-crossing speed in all practices except with the forehand loop-drive in the Two to One practice. LIU, ZHU and CHEN had relatively slower ball speeds. Among them, LIU had slower net-crossing speed with her backhand loop-drive in the Two to One practice; ZHU had slower net-crossing speed with her forehand loop-drive in the Two to One practice and forehand loop-drive against backspin; while CHEN was slower in her forehand moving loop-drive practice.

## References

1. Fang, J. (2003). Research on Table Tennis Collision Using the Simulation Model of the Computer. *Journal of TJIPE*, 18(3): 47-49.
2. Fang, J., Zhang, H., & Yang, J. (2011). Establishment of Simulation System on Throwing Service of Table Tennis. *Journal of Capital Institute of Physical Education*, 23(2): 188-192.
3. Jiang, F., Li, X., & Xu, Q. (2008). Flight Simulation of Table Tennis Ball. *Journal of Qufu Normal University*. 34(1): 104-106.
4. Su, Z., Yu, G., Guo, M., Zhu, L., Yang, J., & He, Z. (2008). Aerodynamic Principles of Table Tennis Loop and Numerical Analysis of Its Flying Route. *China Sport Science*, 28(4):69-71.
5. Wu, H., Qin, Z., Xu, S., & Xi, E. (1988). Experimental Research in Table Tennis Spins. *China Sports Science*, 8(4): 26-32.
6. Yang, C., Yuan, Z., & Liang, Z. (2014). Simulation of Dynamic Characteristics of Table Tennis Rebound. *Computer Simulation*, 31(10): 281-285.

# Table tennis and computer vision: a monocular event classifier

Kevin M. Oldham[1], Paul W. H. Chung[1], Eran A. Edirisinghe[1], Ben. J. Halkon[2]

[1]Department of Computer Science, [2]Wolfson School of Mechanical and Manufacturing Engineering, Loughborough University, Loughborough, LE11 3TU. United Kingdom.

**Abstract.** Detecting events in table tennis using monocular video sequences for match-play officiating is challenging. Here a low-cost monocular video installation generates image sequences and, using the Horn-Schunck Optical Flow algorithm, ball detection and location processing captures sudden changes in the ball's motion. It is demonstrated that each abrupt change corresponds to a distinct event pattern described by its combined velocity, acceleration and bearing. Component motion threshold values are determined from the analysis of a range of table tennis event video sequences. The novel event classifier reviews change in motion data against these thresholds, for use in a rules based officiating decision support system. Experimental results using this method demonstrate an event classification success rate of 95.9%.

**Keywords.** Event classification, table tennis, ball, segmentation, detection, computer vision, optical flow.

## 1 Introduction

Use of computer vision (CV) to support umpire officiating exists in sports with high investments and global audiences, such as football and tennis [1]. It is less popular in low investment sports, or those in which the sophisticated installation requirements are too complex for the environment. One example here is table tennis. Yet the high speed nature of table tennis makes umpire decisions complex and one in which CV would bring benefits. Reducing the data input requirements to a monocular device and interrogating the 2-dimension (2D) ball location coordinates produces sufficient information to detect key match-play events. The limitations of 2D positional data have been mitigated by careful positioning of the video camera [2] and the proposed approach for event analysis is to consider only *changes* in ball motion.

Building on work for event classification using 2D motion in snooker [3], it is proposed that key events in table tennis are given a formal definition based on expected changes in motion combined with their pre and post event states stored in a finite-state machine (FSM). Motion is analysed using a running mean of velocity, acceleration and bearing. It is suggested that the combination of these three characteristics is sufficient for the creation of semantic information required for distinct event identification. This paper provides evidence of the results of such event detection when ap-

© Springer International Publishing Switzerland 2016
P. Chung et al. (eds.), *Proceedings of the 10th International Symposium on Computer Science in Sports (ISCSS)*, Advances in Intelligent Systems and Computing 392, DOI 10.1007/978-3-319-24560-7_4

plied to the data from video sequences, using the Optical Flow (Horn-Schunck) CV algorithm [4]. Results indicate a high degree of success (95.9%) for event detection using a monocular recording.

## 2     The proposed classifier

A continuous assessment of sudden changes in motion triggers an event state check. If the ball's motion matches that of the event description in the FSM then a message is sent to the event engine for evaluation and event state update. To classify events in table tennis, it is proposed that an event is detected by reviewing changes in velocity, acceleration and bearing. The mean ball motion components of velocity, acceleration and bearing are calculated. The temporal motion is compared to the running average of its component motion. If the resulting comparison exceeds the threshold, a confirmation check is triggered followed by the event classification. Upon successful event detection, running averages are reset and the outputs of the previous event become the inputs of the new event. Any missing frame data is not included in the running average, therefore the mean is unaffected by occlusions or non-detection.

### 2.1     Event states

For this proposal an *event* is defined as a significant change in table tennis ball motion from its predicted path, which affects game statistics, scoring or officiating decisions. Only one event can occupy the state of the game at any one time. Any given event can only be preceded and succeeded by a subset of events, as defined by the rules of the game and predictable motion of the ball. Each event, therefore, has a unique definition of both changes in temporal motion data and the preceding and succeeding events. Rule checking is implicit; the predictable sequence of events determines which subsequent event(s) can be valid. If the rule based sequence is broken then the event is a 'fault' and the score (if recorded) is updated. Twenty four distinct event components have been identified for table tennis and these events are grouped into seven top-level events (i.e. table bounce, return, net/let, net/collision, service hit, table edge and over the net), each having specific event motion characteristics.

### 2.2     Event motion thresholds

A description of the combinations of component motion dynamics has been made. These descriptions are placed in the FSM together with valid pre and post event states. The sensitivity of the motion detector must be calibrated to filter motion changes which are not caused by an event. Experimental observations from a range of sequences for each event resulted in a set of distinctive motion change thresholds for each event. The mean measurement of expected minimum change (%) in motion for velocity (min$\Delta v$), acceleration **min$\Delta a$** (min$\Delta a$) and bearing (min$\Delta b$) for each top-level event has been calculated (**Table 1**).

**Table 1.** Event threshold values

| Event | Sample (n) | min$\Delta v$(%) | min$\Delta a$(%) | min$\Delta b(^0)$ |
|---|---|---|---|---|
| Table bounce | 140 | 49 | 0.1% | 58 |
| Net collision (let) | 39 | 9 | 134 | 12 |
| Net collision (net) | 75 | 84 | 160 | 140 |
| Return | 91 | 146 | 189 | 111 |
| Service hit | 38 | 560 | 412 | 205 |
| Table edge | 6 | 24 | 196 | 53 |

As these are the *minimum* values detected during the experiments for the available sample size, an assumption is made that this sample size is sufficient to provide accurate motion threshold values. As motion tends to the limit, detection becomes restricted by available hardware. A number of events require no threshold value in the event test (for example, over-the-net). Instead, these events use current ball location data combined with knowledge of the preceding events in the sequence.

## 3 Classification results

A summary of the results of the novel classifier is presented in **Table 2**. Where events occur in both left-to-right and right-to-left play directions (such as a return) the results have been combined.

**Table 2.** Summary of classification results

| Event | Sample size | %Success | Occlusions | %Success without occlusions |
|---|---|---|---|---|
| Table bounce | 140 | 100 | 0 | 100 |
| Return | 39 | 79 | 6 | 95 |
| Net collision (let) | 75 | 97 | 0 | 97 |
| Net collision (net | 91 | 100 | 0 | 100 |
| Service hit | 38 | 79 | 13 | 95 |
| Table edge | 6 | 67 | 2 | 100 |
| Over the net | 126 | 100 | 0 | 100 |

The results show a promising application of 2D based CV processing to event classification with an overall success rate of 95.9%. The majority of failures occur when the ball, during a return and service[1], is partially occluded by either the player or racket, when using a monocular recording device. Removing occlusion sequences from the 'return' and 'service hit' events improves success rates for both to 95% (giving a total success rate for event detection of 98.8%). As Wong [5] notes in his assessment of ball occlusion during a service 'the view of the ball is blocked and rule

---

[1] There is also the occlusion caused when the ball strikes the *edge* of the table opposite the camera. This is a relatively rare event, and can be easily mitigated by a second camera opposite the first, if considered a requirement.

2.06.04 is violated' suggesting that a failure for the camera to have line of sight of the ball during a serve is equivalent to an umpire not seeing the ball. It is suggested here that with careful positioning of the camera in a location analogous to that of the umpire, this would immediately indicate the service was a fault, prior to any ball detection being made. Table edge events without occlusions were detected with a success rate of 100%, however occlusions occurred in 66% of all edge events. This is primarily due to the camera position only having clear sight of one edge of the table (the nearest length to the camera) and is an issue for the application of a single camera.

# 4    Conclusion

Evidence provided here indicates that 2D data is sufficient for the *majority* of event classifications in table tennis, using derived velocity, acceleration and bearing characteristics. Absolute real-world velocity and acceleration calculations of the ball are not essential for detecting faults or automatic scoring. However, it requires a carefully positioned imaging device. Multiple cameras assist with occlusions, particularly those observed during table edge events but they do not necessarily benefit the detection of a bounce, a let, service or net event. Further investigation is also suggested to evaluate the limits of detection of ball deviation when using a monocular video source.

# References

1. Wang, J. R., & Parameswaran, N. (2004). Survey of sports video analysis: research issues and applications. In *Proceedings of the Pan-Sydney area workshop on Visual information processing* (pp. 87–90). doi:10.1190/66.19
2. Oldham, K. et al., 2015. Experiments in the Application of Computer Vision for Ball and Event Identification in Indoor Sports. In S. Phon-Amnuaisuk & T. W. Au, eds. *Computational Intelligence in Information Systems SE - 27*. Advances in Intelligent Systems and Computing. Springer International Publishing, pp. 275–284. Available at: http://dx.doi.org/10.1007/978-3-319-13153-5_27.
3. Rea, N., Dahyot, R. & Kokaram, A., 2004. Modelling High Level Structure in Sports with Motion Driven HMMS. *ICASSP*, III, pp.621–624.
4. Horn, B. K. P., & Schunck, B. G. (1981). Determining optical flow. *Artificial Intelligence*. doi:10.1016/0004-3702(81)90024-2
5. Wong, P.K.C., 2007. Developing an Intelligent Assistant for Table Tennis Umpires. *First Asia International Conference on Modelling & Simulation (AMS'07)*, pp.340–345. Available at: http://ieeexplore.ieee.org/lpdocs/epic03/wrapper.htm?arnumber=4148683 [Accessed August 12, 2012].

# 3D reconstruction of ball trajectory from a single camera in the ball game

Lejun Shen, Qing Liu, Lin Li, Haipeng Yue

Chengdu Sport University, ChengDu, China

**Abstract.** The 3D ball trajectory provides us quantitative technical or tactical information (e.g. the ball speed of serve). The 3D trajectory can be reconstructed by multiple camera system or single camera system. The single camera 3D reconstruction method is better than the multiple camera one, because it is convenient for the video from television. The existing monocular 3D reconstruction method suffers from the model-drifting problem. We solve this problem using a new cost function. Experimental result shows that our method is more accurate than classical method, because our cost function is a mixture of the physical model and the geometric model.

## 1   Introduction

The 3D reconstruction of moving targets (e.g. ball, players) and static elements (e.g. court, net, or table) of sport competition is important, because some quantitative technical or tactical information can be derived from this 3D reconstruction. For example, the ball speed of serve, the location of players, and the height of ball flying over the net can be computed from the 3D ball trajectory. This information is very useful in the sport coaching process, the feedback of training, the performance analysis etc.

The existing 3D reconstruction methods can be divided into two categories. The first uses multiple cameras to compute the 3D position of ball. The well-known "Hawk Eye" is a typical example. This method requires multiple camera installation, calibration and synchronization. This method is useless in some overseas matches, especially in the field controlled by the competitor, because we *cannot* install our multiple camera system. The limitation of multiple camera method inspires a new 3D reconstruction technology.

The second category of 3D reconstruction method uses a single camera to estimate the 3D trajectory of a ball, named *monocular 3D reconstruction*. This method does NOT require multiple camera installation, which facilitates the usage of this technology. Firstly, some sport video from Television is publicly available on the internet or easily captured from satellite broadcasting without overseas travel. Secondly, we can record the sport match using a single consumer camera, instead of the more expensive multiple camera system. Thirdly, it does not require the permit of the competitor especially when the contest is open or broadcast live. In summary, the monocular 3D reconstruction method can be used in more fields than multiple camera methodology.

© Springer International Publishing Switzerland 2016                                33
P. Chung et al. (eds.), *Proceedings of the 10th International Symposium on Computer Science in Sports (ISCSS)*, Advances in Intelligent Systems and Computing 392, DOI 10.1007/978-3-319-24560-7_5

The monocular 3D reconstruction method requires physical (or geometric) constraints about scene (or target). Reid et al. [5] estimated the 3D position of a ball by a geometric constraint which makes use of shadows on the known ground plane. But shadow is not available in many sport videos. Kim et al. [2] got the heights of a ball according to the player's height, which is not reasonable in table tennis game. Boracchi et al. [2] reconstructed the 3D ball trajectory from a single motion-blur image. Its hidden assumption is not satisfied in the tennis game because the ball is too small in the images. Ohno et al. [4] estimated the ball position by fitting a physical model of ball movement in the 3D space to the observed ball trajectory in the 2D image. Ribnick et al. [6] developed a theoretical analysis and established the minimum conditions for the existence of a unique solution. These two papers [4, 6] are the most related works of this paper, because they do not require special camera setting.

The monocular 3D reconstruction method, however, cannot directly used in the ball trajectory estimation in sport because of "*model drifting*" problem. It means that the estimated ball trajectory is drifting from the true trajectory along the direction which is perpendicular to the camera image plane. We will show the details of this problem later.

This paper presents a new cost function to solve the model drift problem. The intuition behind our solution is that the new cost function improves the reconstruction accuracy by fusing the information both from physical model and from geometric model.

## 2   Methods

### 2.1   Camera Calibration

We use the DLT [1] method and none-linear optimization [3] to get the projection matrix P. Please note that the geometry of sport infrastructure (e.g. the size of court in the tennis competition) is known according to the rule (of International Tennis Federation). Therefore, the tennis court is a calibration object in this paper. Our camera calibration does NOT use any specific calibration object (e.g. the planar pattern [8] or Tsai grids [3]).

According to pinhole camera model [3], P is a $3 \times 4$ projective matrix defined by

$$
\begin{bmatrix} u \\ v \\ 1 \end{bmatrix} = P \begin{bmatrix} X \\ Y \\ Z \\ 1 \end{bmatrix} = \begin{bmatrix} m_{11} & m_{12} & m_{13} & m_{14} \\ m_{21} & m_{22} & m_{23} & m_{24} \\ m_{31} & m_{32} & m_{33} & m_{34} \end{bmatrix} \begin{bmatrix} X \\ Y \\ Z \\ 1 \end{bmatrix} \tag{1}
$$

where $m_{11}, m_{12}, ..., m_{34}$ is the elements of matrix P, $(X, Y, Z)$ is the position in the 3D court model and $(u, v)$ is the observed position in the 2D image. Let $m_{34} = 1$ because the equation (1) uses homogeneous coordinates system. The DLT method needs at least six court-to-image point correspondences between $(X_i, Y_i, Z_i)$ and $(u_i, v_i)$, for i=1,2,...,N. Fig. 1 shows six observed position (four corners of the table and two upper points of net), among which one correspondences is from $(X_1, Y_1, Z_1) = (0, 2.47, 0)$ to $(u_1, v_1) = (482, 190)$.

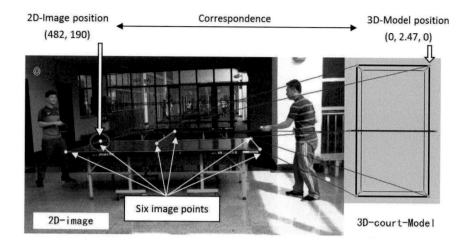

2D-Image position (482, 190)  —— Correspondence ——  3D-Model position (0, 2.47, 0)

2D-image    Six image points    3D-court-Model

**Fig. 1.** The example of six correspondences

The projection matrix P is computed by solving the following linear equations

$$
\begin{bmatrix}
X_1 & Y_1 & Z_1 & 1 & 0 & 0 & 0 & 0 & -u_1X_1 & -u_1Y_1 & -u_1Z_1 \\
0 & 0 & 0 & 0 & X_1 & Y_1 & Z_1 & 1 & -v_1X_1 & -v_1X_1 & -v_1X_1 \\
& & & \cdot & & \cdot & & \cdot & & \\
& & \cdot & & \cdot & & \cdot & & \\
& & & \cdot & & \cdot & & \cdot & & \\
X_N & Y_N & Z_N & 1 & 0 & 0 & 0 & 0 & -u_NX_N & -u_NY_N & -u_NZ_N \\
0 & 0 & 0 & 0 & X_N & Y_N & Z_N & 1 & -v_NX_N & -v_NY_N & -v_NZ_N
\end{bmatrix}
\begin{bmatrix}
m_{11} \\ m_{12} \\ m_{13} \\ m_{14} \\ m_{21} \\ m_{22} \\ m_{23} \\ m_{24} \\ m_{31} \\ m_{32} \\ m_{33}
\end{bmatrix}
=
\begin{bmatrix}
u_1 \\ v_1 \\ \cdot \\ \cdot \\ u_N \\ v_N
\end{bmatrix}
$$

$$(2)$$

where N is the number of court-to-image point correspondences.

## 2.2   3D reconstruction of ball trajectory

We use the Ribnick [6] or Ohno [4] method to estimate the 3D ball trajectory. Firstly, this paper assumes the ball motion is governed by gravity, which is expressed as

$$
\begin{aligned}
X(t) &= X(0) + tV_X(0) \\
Y(t) &= Y(0) + tV_Y(0) \\
Z(t) &= Z(0) + tV_Z(0) - \tfrac{1}{2}gt^2
\end{aligned}
\tag{3}
$$

where $(X(t), Y(t), Z(t))$ and $(V_X(t), V_Y(t), V_Z(t))$ denote the position and the velocity of the ball at time $t$, $g$ denotes the acceleration of gravity ($g = 9.8$ $m/s^2$). Since the physical model (3) is known, the estimated position at time $t$ only depends on the initial position $(X(0), Y(0), Z(0))$ and the initial velocity $(V_X(0), V_Y(0), V_Z(0))$.

Secondly, we estimate the values of initial position and initial velocity $\theta = (X(0), Y(0), Z(0), V_X(0), V_Y(0), V_Z(0))$ which minimize $F_{classic}$

$$\theta^* = \arg\min_{\theta}(F_{classic}) \qquad (4)$$

$F_{classic}$ is the sum of squared difference between the estimated position $(u_P(t), v_P(t))$ and the observed image position $(u(t), v(t))$:

$$F_{classic} = \sum_{t=1}^{M} \left( \{u(t) - u_P(t)\}^2 + \{v(t) - v_P(t)\}^2 \right) \qquad (5)$$

where $(u_P(t), v_P(t))$ is the project position of $(X(t), Y(t), Z(t))$ according to the perspective projection equation (1), and M is the number of observed image points along ball trajectory. A unique solution exists if and only if the $M$ ($M >= 3$) image points are non-collinear [6].

## 2.3    Our improvement

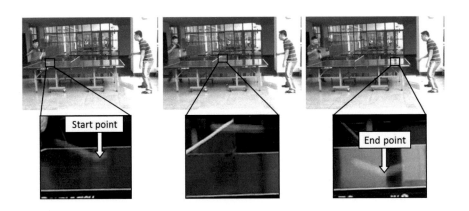

**Fig. 2.** The image sequence of the first stroke in table tennis competition. "Start point" is the observed position where the ball touches the server's court and "End point" is the position where the ball touches the receiver's court.

We show the "model drifting problem" and fix it in this section. Experiment shows that the above Ribnick [6] or Ohno [4] method (hereinafter called classical method) suffers from the model drifting problem. As shown in Fig. 2, the serve (or the first stroke) of table tennis player is recorded using a single consumer camera.

The model drifting problem is shown in Fig. 3, the estimated ball trajectory using classical method (4) is far from the true trajectory. Specifically, the 3D reconstruction error is very large along the direction which perpendicular to the camera image plane.

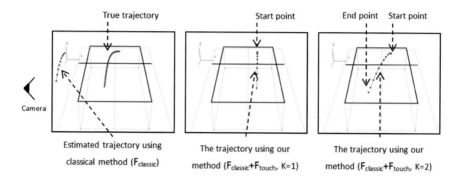

**Fig. 3.** Model drifting problem means that the estimated ball trajectory using classical method is far from true trajectory (1st column). The trajectory using our method with start point is better (2nd column), and the trajectory using our method with both start point and end point is the best result (3rd column).

In order to solve the model-drifting problem, we firstly get the 3D position where the ball touches the court. After rearrangement of (1), we get

$$
\begin{aligned}
(um_{31} - m_{11})X + (um_{32} - m_{12})Y + (um_{33} - m_{13})Z + (um_{34} - m_{14}) = 0 \\
(vm_{31} - m_{21})X + (vm_{32} - m_{22})Y + (vm_{33} - m_{23})Z + (vm_{34} - m_{24}) = 0
\end{aligned} \quad (6)
$$

This is the 3D line equation defined by two planes. Given the observed position $(u, v)$ where the ball touches the court in the 2D image and the projective matrix P, the 3D line from camera to this position is (6). The 3D court plane is also available from the 3D court model if perspective projection matrix P is known. Hence, $(X_C(t), Y_C(t), Z_C(t))$ is the 3D position where the ball touches the court computed by the line-plane intersection at time t.

Secondly, we introduce a new cost function

$$
F_{touch} = \sum_{k=1}^{K} \left( \{X(t) - X_C(t)\}^2 + \{Y(t) - Y_C(t)\}^2 + \{Z(t) - Z_C(t)\}^2 \right) \quad (7)
$$

where $(X(t), Y(t), Z(t))$ denote the position of the ball at time t according to (3), and K is the number of position where the ball touches the table after a stroke. The final cost function is

$$F_{new} = F_{classic} + w \times F_{touch} \tag{8}$$

where $w$ is the mixture weight.

The minimization of cost function $F_{new}$ produces the 3D ball trajectory $\theta^*_{new}$ if $K$ points are available.

$$\theta^*_{new} = \arg\min_{\theta}(F_{new}) \tag{9}$$

Please note that our method (9) equals classical method (4) if $w = 0$. The geometric model is more important than the physical model if $w > 1$. Intuitively, our cost function (8) is a mixture of two models: the physical model (3) and the geometric model (6). The mixture of multiple models is very useful in object tracking [7].

## 3   Experiment

We implement the proposed method using C++ and run it on an Intel Core-i3 3.3GHz PC with 2G RAM. The test videos are recorded from 16 table tennis matches using a consumer camera. We got 3864 ball trajectories and the horizontal ball speed from these videos. The minimal ball speed is 2.11 m/s, and the maximal is 43.79m/s. The percentage of ball speed is shown in Fig. 4.

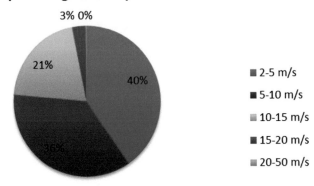

**Fig. 4.** The ball speed of 3864 strokes in table tennis match.

## 4   Conclusion

The geometry of sport infrastructure facilitates the camera calibration, which is more convenient than the multiple camera system. The ball trajectory is estimated by fitting a physical (gravity) model in the 3D space to the ball positions in the 2D image. Moreover, the 3D position where the ball touches the court is computed by the line-plane intersection. This geometric model is more accurate than the physical model in the direction that is perpendicular to the camera image plane. Finally, we solve the model-drifting problem by a new cost function, which fusing the information both from physical model and from geometric model.

This is indeed a first attempt which proved to deserve further investigation. We plan reconstruct the 3D trajectory when the number of position that the ball touches the court is smaller than two (i.e. $K < 2$). The gravity model is reasonable if ball speed is high, but ball spin and atmospheric conditions also affect the ball's trajectory if ball speed is low (e.g. soccer game). We also plan develop a more realistic physical model considering the air friction.

**Acknowledgement.** This work was supported by Scientific Research Fund of SiChuan Provincial Education Department (13ZA0072) and Technology Research and Development Program of Sichuan Province of China (2015JY0148).

## References

1. Abdel-Aziz, Y.: Direct linear transformation from comparator coordinates in close-range photogrammetry. In: ASP Symposium on Close-Range Photogrammetry in Illinois (1971)
2. Boracchi, G., Caglioti, V., Giusti, A.: Ball position and motion reconstruction from blur in a single perspective image. In: International Conference on Image Analysis and Processing (ICIAP). pp. 87–92 (2007)
3. Hartley, R., Zisserman, A.: Multiple view geometry in computer vision, vol. 2. Cambridge Univ Press (2003)
4. Ohno, Y., Miura, J., Shirai, Y.: Tracking players and estimation of the 3d position of a ball in soccer games. In: International Conference on Pattern Recognition (ICPR). vol. 1, pp. 145–148. IEEE (2000)
5. Reid, I., North, A.: 3d trajectories from a single viewpoint using shadows. In: British Machine Vision Conference (BMVC). vol. 50, pp. 51–52 (1998)
6. Ribnick, E., Atev, S., Papanikolopoulos, N.P.: Estimating 3d positions and velocities of projectiles from monocular views. IEEE Transactions on Pattern Analysis and Machine Intelligence 31(5), 938–944 (2009)
7. Shen, L., Liu, Q., You, Z.: Real-time tracking of multiple objects by linear motion and repulsive motion. In: Asian Conference on Computer Vision (ACCV2014). pp. 368–383 (2014)
8. Zhang, Z.: A flexible new technique for camera calibration. IEEE Transactions on Pattern Analysis and Machine Intelligence 22(11), 1330–1334 (2000)

# Part II
# It System for Sport

# Towards a Management Theory for the Introduction of IT Innovations in Top Level Sports

Mina Ghorbani & Martin Lames

Faculty of Health and Sport Sciences, TU München
Georg-Brauchle-Ring, 62; D-80809 Munich, Germany
mina.ghzadeh@gmail.com

**Abstract.** Recent years have seen an increasing number of introduction of IT innovations in top level sports in almost each competitive nation. Practical experiences show that it is a long way with many pitfalls leading from a technological innovation to its successful introduction in sports. The idea of this paper is to make use of management concepts brought forward in economy to provide a conceptual framework for this task. The benefit would be to circumvent common problems, to prepare innovators for typical ones and to improve each stakeholder's understanding of the process. Product and innovation life cycle theory, innovation management theory, and customer relationship management theory are described in the paper. Applicability to the problem under investigation is discussed, and important contributions as well as conceptual differences are pointed out. As examples for best practice, two IT innovations introduced recently in German top level sports are analyzed. National teams (male and female) in the Paralympic sport goalball and Olympic beach-volleyball introduced a computer based match analysis system. One may conclude that it is helpful to borrow conceptual knowledge from economy to better understand the introduction of IT innovations in top level sports.

**Keywords.** IT innovations, top level sports, product life cycle, customer relations management, game analysis systems

## 1    Problem

Introducing innovations in sports is a crucial task for every national top level sports supporting system or organization. The level of competition is very high in Olympic as well as Paralympic sports because of an increasing number of competing nations and the rising standards. Roughly spoken, there are two reasons for success on international level: A perfect training system and the introduction of something new giving a decisive advantage at the moment of competition. As scientific know-how on how to organize training system has become more and more commonplace on international level, many nations are in principle capable to provide optimal support. Thus, the role of introducing innovations will increase in importance for success in competitions. Moreover, it can be noted that the origins of innovations in top level sports lie

© Springer International Publishing Switzerland 2016
P. Chung et al. (eds.), *Proceedings of the 10th International Symposium on Computer Science in Sports (ISCSS)*, Advances in Intelligent Systems and Computing 392, DOI 10.1007/978-3-319-24560-7_6

more and more in the domain of introducing new technologies, especially IT technologies, to improve the quality of work in training and competition.

Managing this process is usually the task of national supporting institutions for to level sports, e.g. AIS in Australia, EIS in England, JIS in Japan, BISp in Germany, and USOC in U.S. In the past, there has been little conceptual knowledge on how to manage this process. Each innovation is more or less treated as individual single case, hardly any management rules are derived systematically from former experiences. Especially in Germany, in these days the system is blamed for low effectiveness failing to bring into work substantial innovations.

As projects with this aim will increase in number it would facilitate and make more effective the work of these institutions if there was conceptual support available. Moreover, a glance at widespread management theories in economy should be helpful because innovations in business play more or less the same role as in sports: they should result in a temporary advantage among competitors.

## 2    Theoretical concepts

### 2.1    Product and innovation life cycle theory

Product life cycle theory describes basically four stages of a product with respect to different management activities that are required to master these stages and to gain maximum profit out of the product over its full life cycle (see fig.1). It was first introduced by Vernon (Vernon & Wells, 1966; Vernon, 1979; Rink & Swan, 1979). In the first phase, introduction, the product is developed and introduced for the first time. The next phase, growth, sees an increasing demand, a more widespread distribution, and, also, other competitors producing their own version of the product. Maturity sees the maximum distribution at lower prices of the product in a very competitive market. Finally, in the decline phase the product becomes increasingly less attractive to the market. It will be replaced by a new, more innovative product.

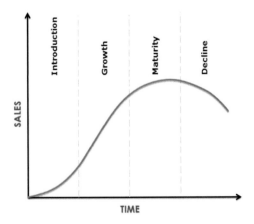

Fig.1: Product Life Cycle Diagram (Komninos, 2002)

Although in top level sports innovations are not made to be sold as often as possible for making profit, we can learn from this concept that there is an initial phase of introducing the product where the costs exceed the momentary profit by far. Only after the market (users) has become accustomed to the use of the product it will develop its full benefits. Most important is the insight, that a new product will lose its innovative characteristics after a certain time. This means that innovations for top-level sports should be treated as a permanent, iterative process rather than being satisfied with one single innovation.

Innovation life cycle theory (Rogers, 1962) addresses the issue of introducing an innovation from the consumer's point of view. It distinguishes between innovators that are people attached to the product developing process, early adopters, early majority, late majority, and laggards. Each innovation has its specific time course and proportion of these consumer categories. What is important for introducing innovations in sports is the insight that one might be among the first people using the innovation by taking big risks of introducing an immature product with a sub-optimal cost-benefit-ratio. Alternatively, one might choose to wait some time before introducing an innovation. Thus, reducing risks of not being effective but increasing the risks of being late compared to competitors.

## 2.2   Innovation management theory

As the introduction of innovations is a most important task in economy, there is a whole branch of management theory that deals with the issue (Biemens, 1992; Trott, 2005). Typically, phase models describe the activities during this process. At the initiate state the problem or the opportunity is identified (see fig. 2). The objectives and available resources have to be defined. In the pilot phase a functional prototype of the innovation is built and tested in practice. In this phase, the innovation needs large support from specialized institutions because it usually operates at a low cost-benefit-ratio initially and in a rather protected environment. On the other hand, failures in the pilot phase may still be interpreted as opportunities for learning. The next phase is the broad implementation of the innovation. Now, full benefits may be drawn from it. Even in this phase, though, improvements of the product are possible and worthwhile because they stem from practical experiences with it in everyday situations. Finally, the product will reach a phase when it becomes obsolete or replaced by a new innovation.

It is easy to imagine that innovation management theory is beneficial in applications for top level sports. An interesting aspect is that at the beginning there might be a problem as well as an opportunity. These are two distinct paths leading to innovations in sports that require different management strategies.

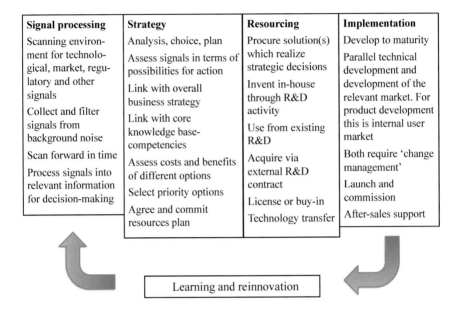

| Signal processing | Strategy | Resourcing | Implementation |
|---|---|---|---|
| Scanning environment for technological, market, regulatory and other signals | Analysis, choice, plan | Procure solution(s) which realize strategic decisions | Develop to maturity |
| | Assess signals in terms of possibilities for action | | Parallel technical development and development of the relevant market. For product development this is internal user market |
| Collect and filter signals from background noise | Link with overall business strategy | Invent in-house through R&D activity | |
| | Link with core knowledge base-competencies | Use from existing R&D | |
| Scan forward in time | Assess costs and benefits of different options | Acquire via external R&D contract | Both require 'change management' |
| Process signals into relevant information for decision-making | Select priority options | | Launch and commission |
| | Agree and commit resources plan | License or buy-in | |
| | | Technology transfer | After-sales support |

Learning and reinnovation

Figure 2: Routines Underlying the Process of Innovation (Tidd, Bessant & Pavitt, 2001)

## 2.3    Customer Relationship Management Theory

A very important and very developed branch of management theory is customer relationship management (CRM; Nguyen & Mutum, 2012; Reinartz, Krafft & Hoyer, 2004). It deals with actions and consequences to be taken with respect to the special customers for a product. Especially the aspect of value-oriented CRM (Evans, Jamal & Foxall, 2006) is very appropriate for acting in sports practice because it puts the customer's values at the focus of its interest. We frequently find different value horizons in sports practice compared to scientific support institutions. CRM provides tools and theories to deal with this important kind of problem in introducing innovations to sports.

Although typical CRM approaches are not applicable to our settings, basic conceptual ideas like reconstructing the values of a customer are a valuable guideline. Especially the area of CRM is not fully exploited yet to provide a better understanding or management rules for IT innovations in sports.

## 3    Examples of Best Practice

In the past decades, several projects were realized in our department that consisted of introducing IT technology to game analysis in German top level sports. In beachvolleyball (Link, 2014) a tool for game analysis was designed and implemented in national adult and youth teams, including the gold medal winners of the London 2012 Olympics (see fig. 3). In Paralympic goalball (Weber & Link, 2014) a similar target was addressed (see fig. 4) and it is interesting to learn from the differences and communalities of the two and other projects.

Most important communalities are the software architecture with database, video interface and a user interface adapted to customer use scenarios. Both systems provide a two-stage mode for analyses: the first one meant for real-time input during the matches, the second one for assessing important aspects that are more difficult to analyze remote and video-based but with in-depth inspection. Interestingly, both systems have image-detection based add-ons to assess typical aspects as economically as possible (goal-ball: ball speed; beach-volleyball: player's positions). Most important differences are the initiation of the project (science staff vs. practice) and the present operating staff (scientists/developers vs. national coaches).

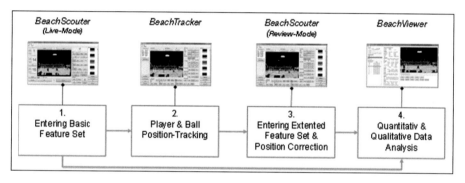

Figure 3: System diagram of Beach Volleyball analysis software

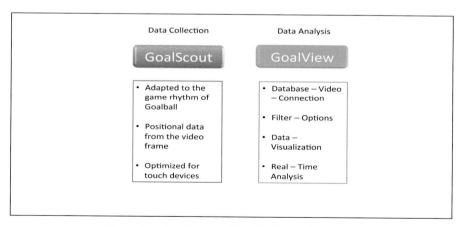

Figure 4: System diagram of Goal ball analysis software

# 4    Conclusions

Although there are considerable differences between introducing innovations to a commercial market and to sports practice, the strategy of borrowing these concepts from management theory seems to be fruitful.

In analogy to economy we propose an adopted phase model consisting of innovation, estimation of potential in sports, pilot study, and implementation as regular scientific service (see fig. 5). Each of these stages has typical main actors, communication networks, criteria of success, and performance standards. Controlling these factors based on the presented concept should help responsible organizations to facilitate the process of introducing IT innovations in top level sports.

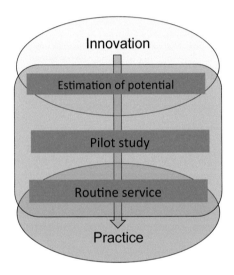

Figure 5: The Life Cycle of Innovation in Sports

# 5    References

1. Biemens, W.G. (1992). *Managing Innovations with Networks*. London: Routledge.
2. Evans, M., Jamal, A. & Foxall, G. (2006). *Consumer behavior*. Hoboken, NJ: Wiley.
3. Komninos I. (2002). Product Life Cycle Management; Urban and Regional Innovation Research Unit, Faculty of Engineering Aristotle University of Thessaloniki; Greece.
4. Link, D. (2014). A Toolset for Beach Volleyball Game Analysis Based on Object Tracking. *International Journal of Computer Science in Sport, 13(1), 24-35.*
5. Nguyen, B. & Mutum, D. S. (2012). A review of customer relationship management: successes, pitfalls and futures. *Business Process Management Journal, 18* (3), 400-419.
6. Reinartz, W., Krafft, M. & Hoyer, W.D. (2004). The Customer Relationship Management Process. *Journal of Marketing Research, 41* (8), 293–305.
7. Rink, D.R. & Swan, J.E. (1979). Product life cycle research: A literature review. *Journal of Business Research, 7* (3), 219-242.
8. Rogers, E. (1962). *Diffusion of innovations*. New York: Free Press.
9. Tidd J., Bessant J. & Pavitt K. (2001), *Managing Innovation,* John Wiley & Sons Ltd, P.53, ISBN: 0-471-49615-4
10. Trott, P. (2005). *Innovation Management and New Product Development*. Prentice Hall.
11. Vernon, R. (1979). *The product cycle hypothesis in a new international environment*. Oxford bulletin of economics and statistics, 41(4), 255-267.
12. Vernon, R. & Wells, L.T. (1966). *International trade and international investment in the product life cycle*. Quarterly Journal of Economics, 81(2), 190-207.
13. Weber, C. & Link, D. (2014). Goalball – Performance Analysis in a Paralympic Sport. In G. Sporis, Z. Milanovic, M. Huges & D. Skegro (Eds.), *ISPAS World Congress of Performance Analysis X , Book of Abstracts (pp. 114-115)*. Zagreb: University of Zagreb.

# Information Systems for Top Level Football

Thomas Blobel & Martin Lames

Technische Universität München, Chair of Training Science and Sports Informatics, Munich, Germany

**Abstract.** In football clubs there are different isolated sources of information. The result is a fundamental need for a central club information system (CIS) for the management. The development of such a system is the purpose of this research. An analysis of the current situation at football clubs has to be done. Based on that, the structure and design for the software solution could be developed. A prototype and a field test will improve the understanding of the needs and habits of the employees at the club. This new information will run into the further development. That iterative process creates a software model that could fulfill the requirements of a central CIS.

**Keywords.** Soccer, information systems, performance analysis, software development

## 1    Introduction

In top level football clubs, typically information is generated by many different sub-systems and sub-organisations. There are different sources of information (such as club-management, public relations, team-management, medical, athletics... etc.) and every field has its own systems to generate and store this information (Lames, 1997). In economy, there are very powerful systems that are developed to generate information out of data from different sources as fundament for decisions of the management. These systems are called management information systems (MIS). This paper presents the idea of developing a MIS for professional football clubs. That would mean that all employees in a club can have access to the data that are relevant for them. Because so little is known, a research cooperation with two youth academies of a first league football club of the German Bundesliga has been started. At the example of the structures in one special field in this youth academy a concept for such a system will be developed. Based on this concept, a prototype will be implemented and observed in a field test. A main target of this work will be the development of deeper analyses across different source systems. That should help to provide the employees at the youth academy with the relevant information independent of the source.

© Springer International Publishing Switzerland 2016
P. Chung et al. (eds.), *Proceedings of the 10th International Symposium on Computer Science in Sports (ISCSS)*, Advances in Intelligent Systems and Computing 392, DOI 10.1007/978-3-319-24560-7_7

Currently there are already existing systems for data collection in football clubs, but they are limited to one or only few fields. A system, that comprehends potential fields of relevance and is able to examine cross-connection between these data, is at the moment not available. The authors are not aware of other approaches at this level of abstraction combining knowledge from sports science and computer science.

## 2 Problem definition

At present, most of the times different data of several sub-systems are stored decentralised at the computers of each employee. Even if there are already some information systems implemented, they are in general limited to their special field and couldn´t be used for the whole club. The consequence is, that there is no central data storage existing, which provides all club internal data. Caused by this, analyses could only be done in the particular field and sub-system. A cross connection between data of different sources is difficult, which means, that an already existing resource couldn´t be fully used. Additionally, barriers exists to access for other employees, because these data are bounded to a special computer or sub-system. Most of this sub-systems follow a strict proprietary logic and need a technical background to understand it. But in football clubs there are many employees without much previous technical knowledge. This increases the barrier additionally. Even the user interface (UI) of these systems doesn´t help to reduce that barrier. Most of that are purely functional and too technical orientated. But the design of the UI could be the key to arouse interest at laymen, help them to understand the logic and make it easier to work with the program.

The aims of this study are to conceptualize a Club Information System (CIS) containing all relevant sources of information, to allow retrievals between them, and meets the demands of the different users in a football club.

## 3 Concept for software modelling

The complete new development of an entire club information system (CIS) would be far too extensive and isn´t the purpose of our project. The primary objective is the conception of a software model with the implementation of a prototype. Therefore it is important to define a sector at a youth academy, for that the software model should be exemplary developed and implemented.

### 3.1 Complex performance diagnostics (CPD)

CPD (Tenga, 2013) has been chosen as the component, where the software model should be developed for first. This is appropriate because there are existing data, the component comprehends different data collection techniques and data can be used immediately at the academy.

| A) Standard Tests | |
|---|---|
| **No. Test** | **Sub-test** |
| 1 linear Sprint | 10m |
| | 20m |
| | 30m |
| 2 Explosive strength | CMJ |
| 3 Agility run | right |
| | left |
| 4 Functional movement screen | 7 tests |
| 5 300-Yard Shuttle Run | |
| 6 lactate test | |
| 7 Body fat | Caliper method |

| B) Additional tests | |
|---|---|
| **No. Test** | **Sub-test** |
| 1 linear sprint | 5m |
| 2 Sprungkraft | SJ left |
| | SJ right |
| | SJ both |
| | CMJ left |
| | CMJ right |
| | Drop Jump left |
| | Drop Jump right |
| | Drop Jump both |
| | Dyn. jump left |
| | Dyn. jump right |
| | Dyn. jump both |
| 3 maximum power | Strength diagnostics |
| 4 Body fat II | infrared |
| 5 Centre of mass | Posturomed |

| C) Explorative Tests | |
|---|---|
| **No. Test** | **Sub-test** |
| 1 3D movement profile | 3D motion capture |
| 2 Perception | Eye tracking |
| 3 Body fat III | Bio impedance anal. |

**Table 1.** Different performance diagnostic tests to generate data for the system.

Standard data in table 1 are already existing. This could be data that are collected at the youth academies at football clubs, like the obligatory DFB performance diagnostic tests (Höner et al., 2013). The additional tests in table 1 focus explosive strength and differential diagnostics of left and rigth foot. The aim is to find dysbalances between both feet and how this effects total jump hight. These additional data are available from 15 players at different age groups. These tests are supplemented by exploratory tests. The sample is limited to three to five players. It´s purpose is to check whether innovative procedures reveal additional relevant information.

## 3.2    Software selection

First, a software was chosen, that served as basic for the development of a prototype (Perl et al., 2002). At the sector of Business Intelligence (BI) in economy there are various software solutions with different specifications available. The following criteria were important for the software selection: Fast and flexible analyses (in memory technology), quick response, flexible user interface (UI) design, quick changes in design and analyses, online applications available and compatible to different data formats. The software that solves these requirements best, turned out to be QlikView. In the companies education program the software is free of charge for universities.

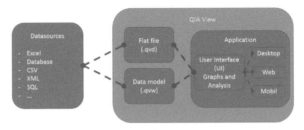

**Fig. 1.** Different steps of data import and software developing with QlikView.

The software can handle major database formats. The loading process uses a database language that is very similar to SQL. After loading the data two data models can be chosen. A flat file or an ER model. This model is saved in QlikView format. That file will be imported into QlikView again and the application can be developed. This separated structure between data model and application provides a very flexible model and the application could be reused again at other clubs.

### 3.3    Data collection

For the software design, data are the most important part. Therefore it was relevant to collect the already existing data from the employees at the academy. After that, these data where edited and normalised. Furthermore they will be added by data of the new performance diagnostic tests and personal data of the players.

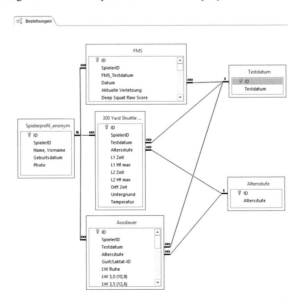

**Fig. 2.** The current data model for the performance diagnostic.

# 4    Software modelling

## 4.1    Status quo at the youth academies

The chief objective at the development is, that the software model should meet the needs of the employees in a football club. Therefore it's important, to detect, how they work, which data are relevant for them and in which data form is needed. This could be found out by personal interviews and a participating observation at the youth academy.

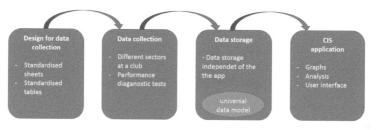

**Fig. 3.** Different steps to develop the software considering the special needs in a football club.

## 4.2    Software development

First, there has been a basic structure for the CIS developed. The system is segmented in different sectors. These sectors are: Homescreen, dashboard, single player, team, complex performance diagnostic (CPD), training, match, video, scouting and medical. At the moment, only the basic structure and the CPD is worked out (see Figure 4). Other parts will follow, to get step by step a complete CPD.

**Fig. 4.** A central information system for the different fields in a football club with the current focus on CIS design and CPD.

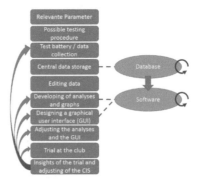

**Fig. 5.** Different steps to develop the CIS with a reachback to the preceding steps. A cycled development at the database and software.

The software development follows an iterative process (Perl & Uthmann, 1997). It makes no sense, to develop the whole software in one step (see Figure 5). So it is important, to develop a prototype, to make a field test, get more information about the habits and wants of the employees and include this into a new prototype.

## 4.3    Developing analyses

For each of these sectors special analyses will be developed. This is one of the most important parts of the software model. Data should be transformed to information and data of different source systems are combined to get more information and relationships between these. This is the part, which delivers new information and a benefit for the employees and the whole youth academy.

## 4.4    User interface (UI)

The UI is not only the screen that shows the information to the users. It is a main part of an intuitive user guidance. For that, it was important, to develop a design, that is attractive to the user, generates interest and helps also to understand the analyses and get through the information (Few, 2006). For the success of such a software tool in the practical environment, the design of the UI and it´s consistency is very important (Lames & Perl, 1997). First there was an overall design concept that is the basis for the whole software model. After that, buttons, boxes and different screens for each topic were developed.

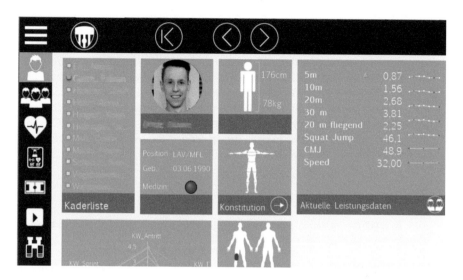

**Fig. 6.** Screen with all summarised data of one player. Additional data and functions could be extended by clicking on the button. By selecting one tile, the user can drill down into data.

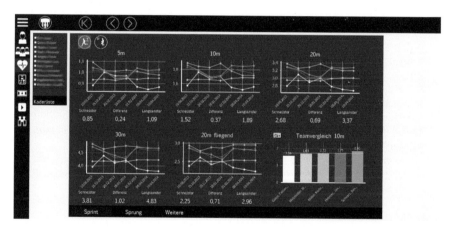

**Fig. 7.** Performance diagnostic of the whole team (currently there are only five datasets available). The user can analyse different players or datasets by selecting them directly in the graphs.

# 5    Conclusions

A CIS can help employees in a football club to get easier access to their relevant data and provides support for their work. Therefore, it is important to have one central system that delivers all information.

However, the success of such a system depends on three factors. Most important are the collected data. They provide the basis for the whole system. The second factor is the CIS by itself. The usability and the scope of performance are the key factors that determine whether the system will be used by the users successfully. At this point it is important to get feedback from the users and adjust and expand the system. The last factor is the user who works with the system. It is necessary that they accept the system and use it for their daily work. If they use these data to get feedback for their work and draw conclusions, a reciprocal relationship between their work, the CIS and the data arises. With the amount of data, the operating time and the experience of the user, it could become a more and more powerful tool for the club.

# References

1. Few, S. (2006). *Information Dashboard Design: The Effective Visual Communication of Data.* Sebastopol: O'Reilly Media Inc.
2. Höner, O., Votteler, A., Schmid, M., Schultz, F. & Roth, K. (2015). Psychometric properties of the motor diagnostics in the German football talent identification and development programme. *Journal of Sports Sciences, 33*(2), 145-159.
3. erLames, M. (1997). Training und Wettkampf. In *Informatik im Sport.* Perl, J., Lames, M. & Miethling, W. Schorndorf: Verlag Karl Hofmann.
4. Lames, M. & Perl, J. (1997). Konzepte für Entwicklung und Einsatz sportinformatischer Werkzeuge. In *Informatik im Sport.* Perl, J., Lames, M. & Miethling, W. Schorndorf: Verlag Karl Hofmann.
5. Perl, J. & Uthmann, T. (1997) Modellbildung. In *Informatik im Sport.* Perl, J., Lames, M. & Miethling, W. Schorndorf: Verlag Karl Hofmann.
6. Perl, J., Lames, M. & Glitsch, U. (2002). *Modellbildung in der Sportwissenschaft.* Schorndorf: Verlag Karl Hofmann.
7. Schumaker, P., Solieman, O. & Hsinchun, C. (2010). *Sports Data Mining.* Heidelberg: Springer.
8. Tenga, A. (2013). Soccer. In *Routledge Handbook of Sports Performance Analysis.* McGarry, T., O'Donoghue, P. & Sampaio, J. Oxfordshire: Routledge.
9. Wiltshire, H. (2013). Sports performance analysis for high performance managers. In *Routledge Handbook of Sports Performance Analysis.* McGarry, T., O'Donoghue, P. & Sampaio, J. Oxfordshire: Routledge.

# Frame by frame playback on the Internet video

## Chikara Miyaji[1]

Japan Institute of Sports Science,
3-15-1 Nishigaoka, kita-ku,
Tokyo, 115-0056, Japan

**Abstract.** Although frame-by-frame is one of the important operation to see sports movement precisely, it is not implemented in the streaming videos on the Internet. This article introduces Smart-method, a new method of frame-by-frame on the streaming videos. It is based on the combination of two methods, streaming and download; streaming is used for playback, the image download is used for frame-by-frame operation, and these operations are switched smoothly, users will not notice these switches. Smart-method is not only enables to see sports movement precisely it opens various applications possible on the Internet; attaching meta-data of the sports videos, flexible thumbnail, and editing (cut and join the videos) on the Internet.

**Key words:** streaming, frame-by-frame playback, and the Internet videos

## 1  Where the problem is?

It was almost 25 years has passed after computers have handled the videos, for example, Apple Inc. introduced QuickTime video technology in the year 1991. Currently no application exists to process videos without computers. In these 25 years, the Internet becomes wide spread and the video format has evolved from SD to HD, and even more to 4k/8k. The videos are shared on the Internet such as YouTube, and are browsed even on smart phones.

Does video on the Internet become easier to handle from these years? Adaptive bitrate streaming2, the most recent technology for the Internet video, only handle play, pause and seek but no frame-by-frame playback. Because of the limitation, video editing still uses local video files. The Internet videos are not comfortable to observe sports movements precisely.

## 2  Why the Internet videos can't perform frame-by-frame?

First of all, it is not designed for these operations, but just for playback videos such as movie clips.

---

[1] chikara.miyaji@jpnsports.go.jp
[2] In this article, the word streaming means Adaptive bitrate streaming.

© Springer International Publishing Switzerland 2016
P. Chung et al. (eds.), *Proceedings of the 10th International Symposium on Computer Science in Sports (ISCSS)*, Advances in Intelligent Systems and Computing 392, DOI 10.1007/978-3-319-24560-7_8

On streaming, the data will not keep on the client after it played back, so it is difficult to perform step backward. Also if seeking will occur on streaming, the time to
seek is not accurate because of the character of streaming. On streaming, the video
file is divided into small fragments, and is downloaded sequentially to playback. Each
fragment starts from the key-frame and followed by the difference frames. So, when
seeking occurs, the client restrains to the key-frame instead of the real frame. This
results that seek can't emulate frame-by-frame playback.

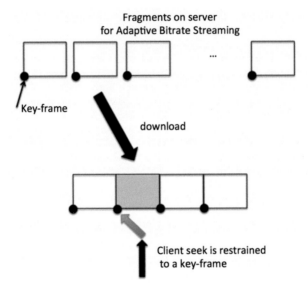

Fig.1: A client seek is restrained to a key frame on adaptive bitrate streaming.

The lack of frame-by-frame is not a lack of a function, but makes many applications impossible to perform on the Internet.

## 3    The purpose of the article

On this article, the author introduces new method of forward and backward frame-by-
frame playback on the Internet, named Smart-method (Streaming Media Application
enRichment Technology method), and explains its implementation. This method is
general to any streaming method, so that it can apply on Apple's HLS or MS's smooth
streaming. After frame-by-frame become possible on the Internet, new applications
become possible. This article introduces new thumbnail methods based on Smart-
method. Also a new video editing on the Internet is explained. It is amazing that
Smart-method expands many possibilities of the Internet videos.

# 4     The implementation of frame-by-frame on Smart-method

Smart-method does not implement frame-by-frame on the streaming mechanism, instead, it implements frame-by-frame by downloading still image of the video form the server. If the still image is downloaded and overlapped on the video itself, it looks like the video player performs a frame-by-frame playback.

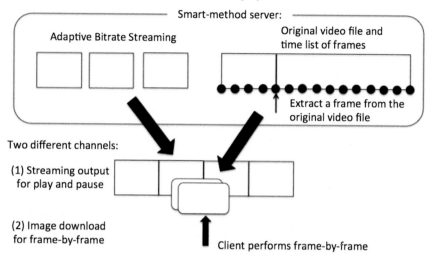

Fig.2: Smart-method uses two channels to play and to perform frame-by-frame.

Smart-method defines APIs to download the still image.
There are two APIs for the server:

## (1) API to get the list of the times (ms) of all frames of the video

http://..server../dir/video/time_list.json => [0, 33, 67, 100, 133, ...]

If the video is high frame-rate (ex. 120fps), the list should be [0, 8, 17, 26,...].
## (2) API to get the image of the frame of a certain time (ms) of the video

http://..server../picture.php?file=dir/video&ms=99 => an image

The server program picture.php, a simple PHP application, returns an image. And inside the program, a ffmpeg command is running:

```
$ ffmpeg –ss 120 -i file.mp4 –f image2 –vframes 30 out%d.jpg
```

This command extracts 30 images of the frames of "file.mp4" start from 120 sec, which is enough to return the image.
If above APIs exist, frame-by-frame playback is done by the sequences below,

1. Pause the streaming video,
2. User choose forward step (or backward),
3. Get the current time of the video in ms,
4. Calculate the time of next frame from API (1),
5. Request the image of the time using API (2),
6. Display the image overlapped on the video

This is exactly what Smart-player (a video player in which Smart-method is implemented)[1] is doing. In the practical situation, it is difficult to set exact time only after pausing, thumbnail search (explain later) will be used before frame-by-frame playback.

A screen shot of Smart-player component on browser

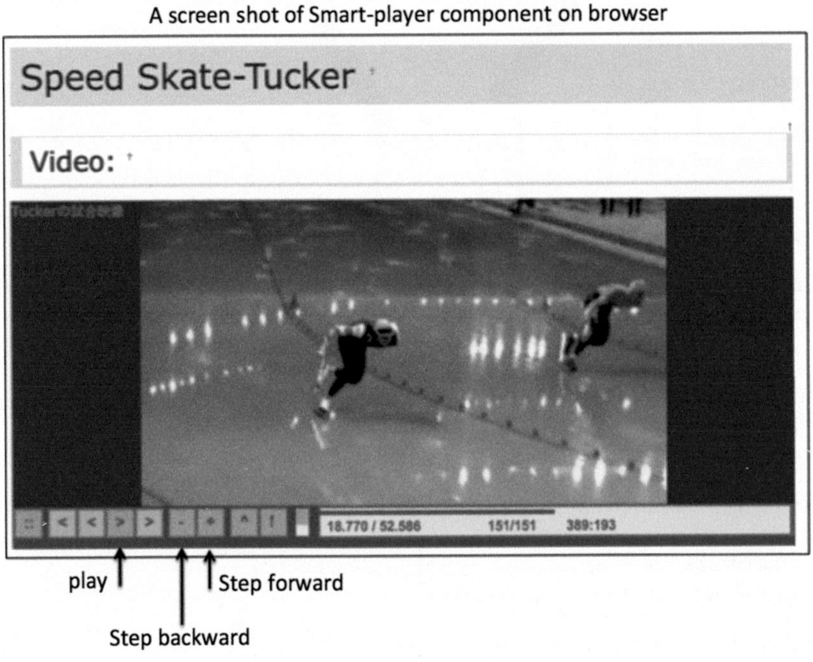

Fig.3: A screen shot of Smart-player component on browser.

The merit of Smart-method is free from the streaming protocols. For example, Smart-player for iOS (Using HLS) and Smart-player for PC (Using Smooth Streaming), both work exactly same on frame-by-frame playback.

The demerit of this method is the requirement of the programs on the server. The existing streaming service such as YouTube can't perform Smart-method.

# 5    Applications of Smart-method

When frame-by-frame become possible, several applications expand its functionalities. Some of the applications explain here.

## 5.1    For Sports video browsing

The original intention to expand video player came form the poor playback of YouTube player. For browsing sports movement, it is very important to see the exact point such as ball release or hitting. Also it is important to use frame-by-frame to understand very fast movement. Although many applications can perform frame-by-frame on local video files, no application performs frame-by-frame on the Internet video. Smart-method helps the user to see the Internet video precisely.

Meta-data is the base of searching the scenes, or analyzing the game. For example, Major League Baseball Advanced Media, one of the top company of sports media industry in US, uses more than 500 peoples for attaching meta-data on their baseball videos. It means that human eyes are important even on the most advanced company. There are two important requirements for attaching meta-data on sports video,

- Meta-data requires exact timing of the event
- Human power is necessary to define the meta-data

Smart-method helps these situations; frame-by-frame helps to find exact time of the event and the use of the Internet helps to share the work with many peoples.

## 5.2    Implementation of thumbnail

For sports video browsing, it is common to see some specific scene instead of watching it from the beginning to the end. Thumbnails, 100 to 200 selected images from the video, are used to find the important scene quickly. For this application, API to download a image is used, and lower image resolution option is added on this API to make download faster. Although Smart-player takes about 7.5 sec to download 150 thumbnails, it will not stop the user operations because it is running as background.

The list of the thumbnail does not need to be laddered, but the time intervals can be varied from sparse to dense. Usually videos of sports will have unimportant scenes and important scenes according to the content, and the thumbnail should reflect these characteristics; unimportant scenes will have sparse intervals, and the important scenes will have dense. This mechanism is called "non-linear thumbnail".

Non-linear thumbnail helps to find the important scenes quickly, but sometimes, each user will have different importance in sports, for example coach's view and judge's view are different. For such case, Smart-player provides "on-demand thumbnail" to download dense thumbnails according to the period where user select.

If thumbnails of all frames are collected in some period, the action of dragging the mouse (for PC) or moving the finger on the screen (for smart phone) makes very smooth movement of thumbnails. Action moving left to right make normal play, and the action moving right to left make backward play, and repeating both actions help to see the movement again and again. This operation is very useful to examine the movement.

The flexibility of the API of Smart-method makes various thumbnail methods possible.

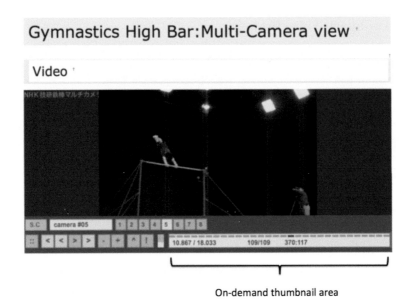

On-demand thumbnail area

Fig.4: A snap shot of using on-demand thumbnail

## 5.3   Implementation for Editing

Streaming uses metafiles to define fragments structure. Based on these metafiles, cut and join are realized by editing of these metafiles. Because of the time period of the fragments, the smallest unit of editing is restricted on its size. For this restriction, frame-by-frame editing based on metadata looks impossible.

Fig.5 shows simple schematic representation of join operation of video A and video B. If user wants to join "part of video A" and "part of video B", the joined metadata will include unnecessary parts.

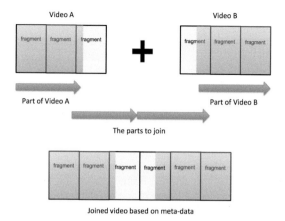

Fig.5: A schematic representation of join operation based on meta-data.

On Fig.6, frame data are shown and blacks are the part to join. For the frame data of video A+B, a joined lists of both black part of video A and video B will be created.

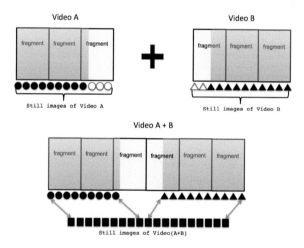

Fig.6: A schematic representation of joined video with frame data.

If user will check the video A+B by frame-by-frame playback, the frame data will show exactly what user edited and will not include unnecessary part. If the user will check the video A+B by replay, unnecessary part will be replayed but that part will be too short to notice.

Here are my assumptions:

- When playing the video, it will not important to include small unnecessary part
- When doing frame-by-frame playback, it is important to see accurate frame editing

Smart-method satisfies these assumptions, and it means that video editing over the Internet is not impossible.

This kind of editing will be useful for simple sports videos editing, such as collect batting scenes or cut some scenes. It will be also useful to use it as proxy editing for professional use, because it includes precise frame-by-frame data.

When 4K/8K videos will become popular, all videos will be uploaded, and editing this way will be realistic.

## 5.4 The possibility for Sports analysis

Almost all current sports video analysis applications will use local video file as the input, because these applications require frame images for their image processing. Smart-method opens the way to get frame image via the Internet. If these applications will include Smart-method mechanism for their input, these applications will have more choices for their input. Also if these applications will be put on the Internet, these applications will become services. This will make more choice for the users.

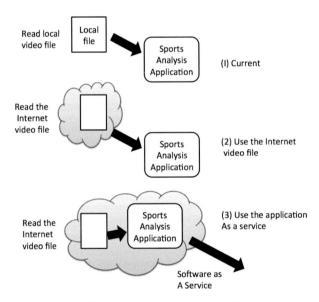

Fig. 7. Three steps to evolve Sports Applications

## References

[1] Chikara MIYAJI: A New Browsing Method for Sports Video on the Internet, Proc. of the 8th International Symposium on Computer Science in Sport (IACSS2011), pp54-57, 2011

# Part III
# Ai in Sport

# Computational system for strategy design and match simulation in team sports

Leonardo Lamas[1], Guilherme Otranto[2], and Junior Barrera[2]

[1] University of Brasilia, Faculty of Physical Education,
Campus Darcy Ribeiro, Brasília, Brazil
[2] University of São Paulo, Institute of Mathematics and Statistics,
Rua do Matão, n.1010, Cidade Universitária, São Paulo, Brazil

**Abstract.** The goal of the present work was to create a computational system that supports the design of strategies and the match simulation based on those strategies. A formal model of team strategy and match dynamics supported the specification of the computational system. In this model, team strategy was defined as a discrete dynamic system. The specification of individual action rules enables the team players to organize the collective action in every state of the system. A play is modelled by a sequence of compatible pairs of states. The system implementation encompasses a designing tool, whose resultant strategies are used as the input in a simulator capable of recognizing match states and applying the defined strategy to plan actions. Besides the inherent contribution to the investigation of team sports performance features, the presented framework may be helpful in other scientific areas, such as those that investigate cooperative actions in competitive environments and to the design of video-games with a greater realism, approximating them to real match simulators.
Key-words: dynamical system, action rule, planning, artificial intelligence

## 1 Introduction

In team sports, the set of specifications conceived by the coaching staff to support the collective action of the players define the concept of strategy [3]. Even though in many circumstances of the match the execution of the team strategy should be modified due to the spatio-temporal constraints mutually imposed by the adversaries, it greatly influences the tactical patterns observed in the confront [5].

The empirical observation of a match leads to the conjecture that, in most of the cases, the more collectively a team plays the greater should be its performance. Hence, the success of a team appears to be related both to the features of the designed strategy and to the team efficiency to execute them in the match. Computational systems that support the design of team strategies and simulate their execution may contribute to the scientific investigation of the several elements involved in this complex context. However, in the present moment, interactive computational environments for strategy design (i.e. environments that can correct and improve the human planning) and team sports simulators, analogous to

© Springer International Publishing Switzerland 2016
P. Chung et al. (eds.), *Proceedings of the 10th International Symposium
on Computer Science in Sports (ISCSS)*, Advances in Intelligent Systems
and Computing 392, DOI 10.1007/978-3-319-24560-7_9

those in existent for improving car races' strategies [6] and for training the pilots in the designed strategies, are not available. Interestingly, in computer science, several algorithms have been systematically improved to reproduce more precisely the competition features of team sports in video games [7, 2], indicating an eminent common research field for scientists from this area and sport sciences. Based on this inter-disciplinary approach, the goal of the present work was to create a computational system that supports the design of strategies and, based on the designed strategies, simulate the match in team sports.

## 2   Methods

### 2.1   Specification of the computational system

A model of the team strategy and the match dynamics [4] supported the development of both modules of the computational system, the team strategy designer (TSD) and the match simulator.

The team strategy model was defined through the following elements: i) control of a player actions by strategy specifications; ii) organization of cooperative relations between team players in each strategy state; iii) graph representation of sequences of states [4]. Additionally, strategies of the adversary teams were inputs to the match dynamics [4].

The most fundamental element of the model is the action rule, a conditional statement of the form if <condition>, then <action> (e.g. in basketball defense, if <an attacker overcomes my teammate>, then <I should help defending his attacker>). An action rule formalizes the logical control of the players dynamics through the action choice of the players in a given context. Players' action rules provide the specifications for creating small cooperative groups of team players, defined as strategic units (SUs), which should perform a coordinated action to achieve a certain strategic goal in a state.

A strategy state is composed by the following constitutive elements: i) players in the match field; ii) a region for each player, represented by an equivalence class for positioning that encompasses all points in the match field with similar meaning for a player in a given state; iii) players dynamics; iv) whether the team has ball possession (i.e. offense or defense); v) the ball dynamics. A state transition specifies the roles of the team players (i.e. trajectory in the match field and respective technical skills performed) and their dynamics. To design a consistent sequence of states, two subsequent states must be compatible. Thus, the output of the first state should be identical to the input of the second state. A team strategy is composed of a finite set of states and connections between pairs of compatible states. A play is modelled by a sequence of compatible pairs of states and two or more plays may be originated in a common state. Thus, every team strategy can be represented by a directed graph. The graph enables the identification of the general structure of the strategy and provides easy visualization of specific sequences of states (i.e. plays) in the context of the complete strategy.

In a match, during the interval between two subsequent interruptions, the confrontation is continuous and both teams try to make adjustments on their performance to approximate to the specifications of the strategy states. Hence, each modification that emerges from the confront, originated from the interaction between the two teams, generates information that is confronted through feedback, with the planned state from the team strategy. This input leads to the comparison between the match state and the closest state specified in the strategy of each of the teams. Based on the result of the comparison, the next goal of the strategy is defined, considering the alternatives presented by the team strategy (Figure 1).

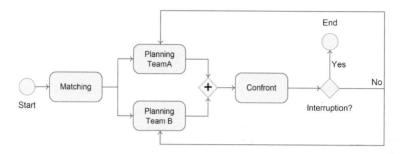

**Fig. 1.** Match dynamics, where: circles represent the start and the end of the flux; rectangles represent processes; diamonds with a cross and empty diamonds represent, respectively, a flux integration and a Boolean test; arrows indicate the direction of the control signal.

## 2.2   System design

The framework of the computational system was divided into two separate modules, the team strategy designer (TSD) and the match simulator. The TSD module allows an user to specify a team's strategy, according to the formalism of the strategy model (see Section 2.1), and creates a file to represent it. The designed strategies for two teams can also be loaded into the match simulator to create a virtual match. The simulation cycle is composed of 3 main processes, the collective planning, the individual interference and a confront. A confront is simulated until the calculated plans cease to be valid, for instance, if a crucial action fails or ends, triggering another round of planning.

The high level collective planner uses the team strategy and the current match state as input. The objective of the planner is to find a path in the team strategy that is both applicable in the current situation and beneficial to the team if successfully navigated. The plan is then modified on an individual scale to introduce automatic individual behaviours (e.g., players can adjust their speed of displacement based on level of fatigue). The planned actions are then simulated until the next round of planning becomes necessary. This cycle is illustrated in Figure 2.

The framework of the computational system can be customized for different team sports. In the present work, the applications were made in the context of basketball.

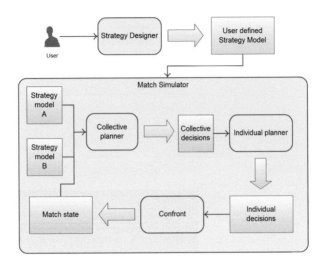

**Fig. 2.** System overview and match simulator main cycle

## 3    Results

### 3.1    Simulation input

The TSD generates a XML file output with the complete description of the specified strategy. Also in the TSD environment, the XML file is represented through a graph (Figure 3A), where the nodes are strategy states (Figures 3B and 3D) and the edges are states transitions (Figure 3C). Then, it provides immediate visualization of the general structure of the designed strategy. In a state, the TSD supports the specification of the players label, players' areas (i.e., equivalence classes for positioning), and ball possession (a small orange circle represents the ball). In the case of the defensive players, their body rotations are also considered due to its relevance for defensive displacements (Figures 3B and 3D). For the purpose of the simulation, two previously specified strategies should be available.

### 3.2    Planning, Matching and Simulation

In the TSD, the user can describe complex strategies in a small state space due to its use of equivalence classes. This allows the description of many similar

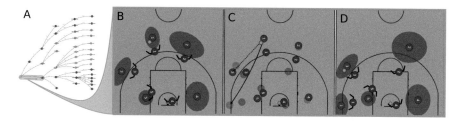

**Fig. 3.** Top-down visualization of a team strategy: 3A - graph of a complete strategy; 3B - previous state; 3C - states transition; 3D - subsequent state. Outer orange ellipses indicate equivalence classes for positioning.

situations in a single strategy state and greatly reduces the strategy graph explored when a plan is calculated. Figure 4 presents an integrative perspective of planning, matching and simulation.

The actual planning problem can be solved using traditional planning algorithms such as the Real-time dynamic programming (RTDP)[1] with one major distinction: the state representation. The planning problem is defined as follows:

1. A state space $S$: the nodes of the strategy graph;

2. An initial state $S_0 \in S$: the node returned by a matching algorithm;

3. A subset of goal states $G \subseteq S$: the terminal states on the graph. A terminal state is any state where a player is in the position to finish the play (i.e., score);

4. A set of actions $A$, and a transition function $f(s, a) \to S$, with $s \in S$ and $a \in A(s)$: the edges of the graph;

5. A function to assign cost to the actions $c(a, s) \in R^+$. This function represents the risk of a transition and is currently assigned by the user when generating the strategy graph.

A new matching layer must be added to the classical planning algorithms (RTDP) to account for the gap between a match state and the states available in the team strategy. The new layer assigns a match state to the strategy state that best represents it. This assignment can be a perfect fit or an approximation.

A metric to compare a match state to a strategy state was devised to allow the matching layer to function regardless of the match state. Thus, the layer will always return a matched strategy state, even if there is no perfect fit (Figure 5). The metric currently in use is the sum of the quadratic distances between all players in the match state and their counterparts in the strategy state. Since many allocations of players can be used, the one that minimizes the sum is chosen.

The metric considers the distance between a match state player and his counterpart in the strategy state to be zero when the match player is inside the equivalence class for positioning of his counterpart (e.g., players A, C and E in Figure 5). Otherwise the distance considered is the squared Euclidean distance between his position and the closest point to it in the equivalence class for positioning (e.g., players D and B in Figure 5).

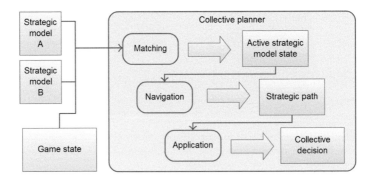

**Fig. 4.** The collective planner execution: Matching provides an initial state; Navigation implements a classical planner; Application adapts the plan to the current match state.


The correct strategy state is then calculated by locating the state which minimizes the sum of the metrics for all players. This calculation also yields the allocation of each player (i.e., his strategy state counterpart).

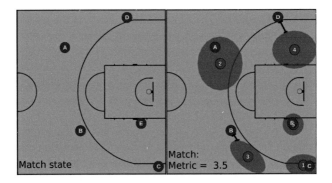

**Fig. 5.** A match state (left), with players assigned in blue, and its counterpart in the strategy (right), with players assigned in orange. The metric 3.5 is the sum of the square distances between players A to E positioning in the match state and their counterparts' equivalence classes for positioning in the strategy state.


The planner extracts a high-level course of action (i.e., a path on the strategy model) using the matched state as an initial node. This is an approximation of the ideal solution, which means it is compatible but not an exact fit for the current match state.

An application process is used to adapt the plan to a specific match state by setting parameters to the actions to be executed. These actions are simple behaviours used by the simulation to execute a plan. An action describes a behaviour (e.g., run, pass, throw, steal) that can be parametrized to be performed exactly as anticipated by the plan.

Once a plan is calculated and adapted, the simulator creates, assigns and executes actions for each player. Some actions have a probabilistic result that can be randomized during the execution (e.g., a score attempt, pass). If one of these action fails, the plan is discarded and another round of planning is triggered. Figure 6 illustrates the simulation environment.

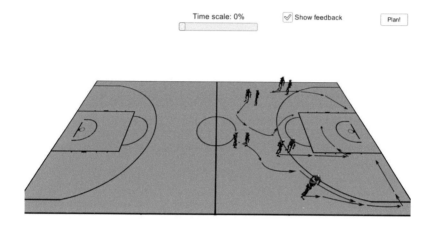

**Fig. 6.** Simulation environment: offensive team (black), defensive team (blue), arrows on the court indicate players' trajectories.

## 4    Conclusion

The main contribution of this study was to develop a computational system that enables the design of team strategies and uses these strategies as input for simulating the match in team sports. This system may support the enhancement of the knowledge about team sports through testing possible features of efficient strategies design and respective consequences for real matches through simulation procedures. This framework may be helpful for coaches and players to improve their capacity to plan the collective performance, interpret the opposition constraints and act more efficiently. Both strategies design and simulation procedures have been conducted in the context of basketball. Nonetheless, the coverage of the model that provides theoretical support for the system enables its customization to a large range of team sports.

These results may have positive impact in other scientific areas as well. For instance, it is possible that the planning and matching structure based on a model for representing strategies and the match dynamics contribute to the design of video-games with a greater realism than the observed in the present games, approximating them to real match simulators. Additionally, it may contribute to solve problems in some branches of artificial intelligence, such as modelling of the collective action in competitive environments, for instance in robot soccer tournaments.

**Acknowledgments**: LL was supported by Fundacão de Amparo à Pesquisa do Distrito Federal. JB is supported by Conselho Nacional de Pesquisa, grant: 306442/2011-6.

# References

1. Bonet, B., Geffner, H.: Labeled RTDP: Improving the convergence of real-time dynamic programming. In: Proceedings of the Thirtheenth International Conference on Automated Planning and Scheduling (2003)
2. Chan, B., Denzinger, J., Gates, D., Loose, K.: Evolutionary behavior testing of commercial computer games. In: Proceedings of the Congress on Evolutionary Computation (2004)
3. Grehaigne, J., Godbout, P., Bouthier, D.: The foundations of tactics and strategy in team sports. Journal of Teaching in Physical Education 18, 159–174 (1999)
4. Lamas, L., Barrera, J., Otranto, G., Ugrinowitsch, C.: Invasionteamsports: strategy and match modeling. International Journal of Performance Analysis in Sports 14, 307–329 (2014)
5. Lamas, L., Santana, F., Otranto, G., Barrera, J.: Inference of team sports strategies based on a library of states: application to basketball. In: Proceedings of the 2014 KDD Workshop on Large-Scale Sports Analytics (2014)
6. Wloch, K., Bentley, P.: Optimising the performance of a formula one car using a genetic algorithm. In: Proceedings of the 8th International Conference of Parallel Problem Solving from Nature (2004)
7. Xiao, G., Southey, F., Holte, R.: Software testing by active learning for commercial games. In: Proceedings of the Congress of the American Association for Artificial Intelligence (2005)

# Soccer analyses by means of artificial neural networks, automatic pass recognition and Voronoi-cells: An approach of measuring tactical success.

Jürgen Perl, University of Mainz

Daniel Memmert, German Sports University Cologne

**Abstract.** Success in a soccer match is usually measured by goals. However, in order to yield goals, successful tactical pre-processing is necessary.

If analyzing a match with the focus on "success", promising tactical activities including vertical passes with control win in the opponent's penalty area have to be the focus. Whether or not a pass is able to crack the opponent's defence depends on the tactical formations of both the opponent's defence and the own offence group.

The methodical part of the contribution consists of three steps:

(1) The first step describes how to analyze the formations of tactical groups by means of an artificial neural network, which is integrated in the *DyCoN*-tool. I.e. the positions of the players are condensed to those of tactical groups, and the formations of the tactical groups are mapped to a small number of characteristic patterns. In this way, the teams' activities can be reduced to interactions of tactical patterns, making it much easier to automatically detect regular and/or striking tactical features. Those tactical features build the context for measuring success, as described in the following steps:

(2) Successful passes from a passing to a receiving player of a team are necessary preconditions for opening or continuing successful attacks. The software *SOCCER* is able to automatically calculate those passes based on the position data of the players and the ball. Along with the information from (1) it can be recognized, which types of formation interaction are helpful for generating successful or "dangerous" passes.

(3) Up to this point, "successful" just means that the receiving player effectively achieves control of the ball. Based on the Voronoi-approach, the pass is seen as more successful and actually "dangerous", if it causes a higher rate of spatial control in the opponent's penalty area. The *Voronoi*-tool calculates tactical space control and therefore helps to measure the tactical success of a pass.

The combination of the described steps of automatic position-based analyses can be helpful for a deeper understanding of match dynamics and measuring tactical success.

The tools *DyCoN*, *SOCCER* and *Voronoi* have been developed by J. Perl, University of Mainz, in cooperation with D. Memmert, DSHS Cologne.

© Springer International Publishing Switzerland 2016
P. Chung et al. (eds.), *Proceedings of the 10th International Symposium on Computer Science in Sports (ISCSS)*, Advances in Intelligent Systems and Computing 392, DOI 10.1007/978-3-319-24560-7_10

## 1    Introduction

The basic idea of SOCCER (Perl & Memmert, 2011; Grunz, Memmert & Perl, 2012) is to reduce the complex match dynamics to the main tactical states and transitions. This approach is adopted from technical process analysis in which proceeding steps are defined by context-depending transitions of states, based on current states and conditions. Doing it this way, a process can not only be planned, described and analyzed but also be simulated for tactical optimization.

Since the late 1970s, it has been the idea of one of the authors (Perl) to apply this approach to sport games and in fact he was successful for instance in applying this to tennis. The main problem, however, was the lack of data. Particularly in case of soccer the process dynamics is highly based on the positions of the players and the ball – and all analysis approaches remained to be academic as long as such data was not easily available.

Since about 2010 the situation has changed a lot due to new position data recording technologies. Achieving information from that data turned out to be the main problem now.

In a first step, trivial information like speed or running distances of the players can be calculated which, however, does not help that much in order to understand tactical concepts.

In the SOCCER-project the positions of the players are condensed to those of tactical groups. The time-depending sets of the players' positions of a tactical group is analyzed by an artificial neural network and thereby mapped to a small number of characteristic patterns which are called formations. In this way the complex interaction of the players can be reduced to a much easier interaction of the regarding formations. On this first level, analysis is still restricted to frequencies and distributions like coincidences of defence formations of team A against offence formations of team B. Using that information on formation coincidences, frequent pairs of defence-offence-formation can be taken to define the states of the playing process, where then the transitions are defined by changes of those transition-pairs. This approach makes sense, because experience shows that such formation-pairs are normally stable over phases of about 5 seconds.

Now the following question has to be answered: Under which contextual conditions do transitions occur – i.e. what events cause the changes of formation-pairs?

Of course there are a lot of possible reasons. In the SOCCER-project we concentrate on passes because passes change the situation of a match and thus can force the tactical groups to react by changing their formations.

Therefore, on the second level of analysis, the passes are taken from the position data and the frequencies of transitions of the type "formation-pair – pass – formation-pair" are calculated.

The question after collecting this data is whether or not such a transition was tactically successful. I.e. the result of a pass has to be measured in terms of tactics.

Although goals on the one hand with no doubt are the most important and most valuable events in soccer they have one important disadvantage: they are comparably rare and – more or less – stochastically distributed over the sequence of tactical activities.

On the other hand space control seems to be necessary in order to generate pressure and prepare dangerous situations – even if sometimes seemingly harmless long passes can result in unexpected goals.

Therefore, on the third level of analysis, the passes have to be analyzed regarding the space control they generate in the critical areas of the opponent team.

Altogether, these three steps allow for replacing the abstract scheme

"situation i    →    event + success    →    situation j"

by the concrete scheme

"formation-pair i  →  pass + space control  →  formation-pair j".

In the following, the three analysis steps mentioned above are described in more detail. All results are achieved from the same exemplarily analyzed match.

## 2    Typing of formations by means of *DyCoN*

Tactical groups like offence or defence form shapes that change depending with situations and interactions. In order to detect tactical behaviour behind those changes it is helpful to reduce the big number of those shapes to a small number of characteristic types or patterns and study their interaction dynamics. Artificial neural networks of type Kohonen Feature Map (KFM) are able to learn types from fed examples, as is shown in figure 1. DyCoN, as an advanced development based on KFM, is able to do so with only a small number of examples and to continue the learning process continuously (KFM, DyCoN: see Perl et al. (2013) and Memmert & Perl (2009)).

More information about net-based soccer analysis can be found in Lees, Barton & Kershaw (2003) and Leser (2006).

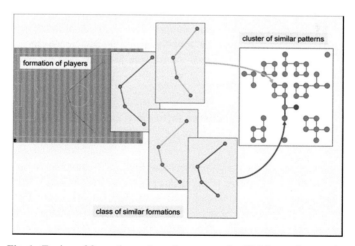

**Fig. 1.** Typing of formation-patterns by means of artificial neural networks.

The results of that net-based typing are prototypes of formations, which in the following are simply called formations. Some examples can be seen on top of figure 2 together with their frequencies. The matrix on the bottom right gives information about formation-pairs - for example: the pair of defence formation 2 of team A and offence formation 8 of team A has a frequency of 23. On the bottom left one situation of "A2 against B8" is shown on the playing field.

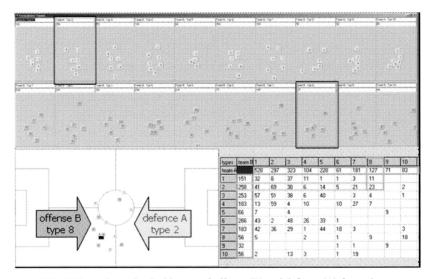

**Fig. 2.** Examples and coincidences of offence (B) and defence (A) formations.

A formation of one of the teams is not only a reaction on the formation of the other team, but also represents a tactical idea depending on the actual match situation, and therefore the formation is expected to change with the changing situation (also see Perl, Grunz & Memmert (2013)). One important type of situation-changing event obviously is the pass, in particular the long vertical one, directed to the opponent's defence zone. Therefore in the next step the relation between passes and formation-pairs is analyzed.

## 3    Recognizing the relation between formations and passes by means of SOCCER

One of the components of the game-analysis-tool SOCCER calculates a match proto-col based on position data, where in particular passes are recognized. The matrix in figure 3 gives a complete overview on the passes of team B in the context of the relat-ed formation-pairs. It shows for example that there was no pass related to formation-pair "A2 against B8" from figure 2. The greatest number of 9 passes is given for the formation-pair "A6 against B5".

| | B | 1 | 2 | 3 | 4 | 5 | 6 | 7 | 8 | 9 | 10 | 11 | 12 | 13 | 14 | 15 | 16 | 17 | 18 | 19 | 20 |
|---|---|---|---|---|---|---|---|---|---|---|----|----|----|----|----|----|----|----|----|----|----|
| A | 163 | 24 | 16 | 17 | 10 | 25 | 4 | 14 | 5 | 2 | 3 | 7 | 6 | 4 | 1 | 4 | 2 | 4 | 2 | 7 | 2 |
| 1 | 12 | 1 | 4 | 2 | 1 | | | | | | | | 1 | | | 1 | 2 | | | | |
| 2 | 1 | 1 | | | | | | | | | | | | | | | | | | | |
| 3 | 12 | 4 | 1 | 1 | | 3 | | | 1 | | | | 2 | | | | | | | | |
| 4 | 22 | 2 | 5 | | 5 | | 2 | 3 | 2 | | | | 2 | | | | | | 1 | | |
| 5 | 4 | | | 2 | | | | | | | | | | | 1 | | | | | | |
| 6 | 19 | 2 | | 3 | | 9 | | | | | | | 2 | 1 | | | 2 | | | | |
| 7 | 20 | 3 | 1 | 5 | | 6 | 2 | 2 | | | | | | | | | | | 1 | | |
| 8 | 2 | 1 | | | | | | | | | | | | | | | | | 1 | | |
| 9 | 4 | | | | | | | 1 | | 1 | | | 1 | | | 1 | | | | | |
| 10 | 6 | | | | 1 | | | 2 | | | | | | | | | 2 | | | | |
| 11 | 7 | | 2 | | | | | 1 | | | | 1 | | | | | | | | 3 | |
| 12 | | | | | | | | | | | | | | | | | | | | | |
| 13 | 5 | 1 | | 1 | | 2 | | | | | 1 | | | | | | | | | | |
| 14 | | | | | | | | | | | | | | | | | | | | | |
| 15 | 7 | 3 | | 1 | | 1 | | 1 | | | | | | | | | | | | | 1 |
| 16 | 3 | 1 | | | | | | | | | | | | | | | | | | | |
| 17 | | | | | | | | | | | | | | | | | | | | | |
| 18 | 5 | 3 | | | | 1 | | | | | | | | 1 | | | | | | | |
| 19 | 4 | 2 | 1 | | | | | 1 | | | | | | | | | | | | | |
| 20 | 2 | | | | 2 | | | | | | | | | | | | | | | | |

Fig. 3. Numbers of passes of team B correlated to the formation-pairs.

Based on that information, it can be analyzed how passes change situations – i.e. how a starting formation-pair is transferred to a succeeding formation-pair during the pass.

The matrix in figure 4 shows the pass-correlated situation-changes. Left and on top are the relevant formation-pairs. The entries are the numbers of transitions caused by passes of B.

As an example, the most frequent situation (A6,B5) in 6 of the 9 passes of B stays unchanged, in 2 cases B changes its formation to B1 resp. B3, and in one case A changes its formation to A7.

| A | | 1 | 1 | 2 | 2 | 2 | 3 | 3 | 3 | 3 | 3 | 4 | 4 | 6 | 6 | 6 | 6 | 7 | 7 | 7 | 7 |
|---|---|---|---|---|---|---|---|---|---|---|----|---|---|---|---|---|----|---|---|---|---|
| B | | 1 | 3 | 1 | 2 | 3 | 1 | 2 | 3 | 5 | 12 | 2 | 7 | 1 | 3 | 5 | 12 | 1 | 2 | 3 | 5 |
| 1 | 1 | | | | | | | | | | | | | 1 | | | | | | | |
| 1 | 3 | | 1 | | 1 | | | | | | | | | | | | | | | 1 | |
| 2 | 1 | | | | | | | | | | | | | | | | | | | | |
| 2 | 2 | | | | | | | | | | | | | | | | | | | | |
| 2 | 3 | | | | | | | | | | | | | | | | | | | | |
| 3 | 1 | | | | | | | | 3 | | | | | | | | | | | | |
| 3 | 2 | | | | | | | | | 1 | | | | | | | | | | | |
| 3 | 3 | | | | | | | | | | | | | | | | | | | | |
| 3 | 5 | | | | | | | 1 | 2 | | | | | | | | | | | | |
| 3 | 12 | | | | | | | | | 1 | | | | | | | 1 | | | | |
| 4 | 2 | | | | | | | 1 | | 1 | | 1 | 1 | | | | | | | | |
| 4 | 7 | | | | | | | | | | | | 2 | | | | | | | | |
| 6 | 1 | | | | | | 1 | | | | | | | | | | | | | | |
| 6 | 3 | | | | | 1 | | | | | | | | | 1 | | | | | | |
| 6 | 5 | | | | | | | | | | | | | 1 | 1 | 6 | | | | | 1 |
| 6 | 12 | | | | | | | | | 1 | | | | | | | | | | | |
| 7 | 1 | | | | | | | | | | | | | | | | | | 2 | | |
| 7 | 2 | | | | | | | | | | | | | | | | | | | 1 | |
| 7 | 3 | | | | | | | | | | | | | | | | | | 1 | 3 | |
| 7 | 5 | | | | | | | | | | 1 | | | | | | | 1 | 1 | | 3 |

Fig. 4. Number of transitions between formations-pairs during passes of team B.

Obviously, most of the entries are in the diagonal – meaning that the corresponding passes do not change those formation-pairs. On the first glance it might be confusing that theses passes seem to rarely change the situations. The reason is that the underlying tactical groups are often moving their positions as a whole without changing their players' position relative to each other. With a long vertical pass, for example, the players of the attacking group move vertically towards the opponent's goal without significantly changing the geometric relations between players' positions, while the opponents' defence group is retarding in a similar way.

The chance of getting a goal can be improved by enlarging the pressure in critical areas close to the opponent's goal. The corresponding activities are passes together with formation changes. Therefore the rate of space control can be taken to measure the success of passes respectively of the corresponding transitions between formation-pairs.

To this aim it is necessary to achieve information about the space control that was generated by the pass and the corresponding changes of players' positions.

## 4    Voronoi cells as context information and success measure

The Voronoi-cell of a player is the area of all points on the playing field he can reach faster than any other player – under the idealistic assumptions of identical reaction and speed. Nevertheless, if not taken as an individual property but summed up over the players of a tactical group it measures the group's space control rather precisely. In the presented approach two critical areas are in the focus: The penalty-area and the 30-m-area (green vertical line in figure 5) of the opponent's half. At each point in time the percentage rate of space control of the offence group in these areas is calculated by the Voronoi-component of SOCCER. Figure 5 shows an example of a high rate of control, which the blue attacking team has reached in both areas.

More information about Voronoi-cells in sports and soccer can be found in Fonseca et al. (2012) and Kim (2004).

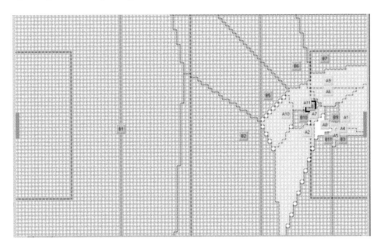

**Fig. 5.** Space control of attacking team B (blue) against defending team A (yellow).

The rates of space control now can be combined with the corresponding changes of situations, i.e. transitions of formation-pairs, as is shown in figure 6. The transitions with comparably high average rates (>10%) are marked by colours:

Blue colour highlights the cases in which relevant space control is given if the formation-pair is unchanged or the attacking team B changes its formation. Example: (A6,B5) is unchanged or is changed to (A6,B1) or (A6,B3).

Yellow colour highlights the cases in which relevant space control is given if only the defending team A changes its formation. Example: (A6,B3) is changed to (A2,B3).

Violet colour highlights the cases where relevant space control is given if the formation-pair is changed by A as well as by B. Example: (A4,B2) is changed to (A3,B5).

Further it can be seen that two of the blue areas characterize groups of transitions where B varies the response to the defence formations of A.

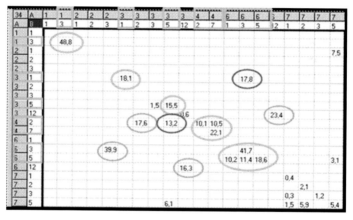

**Fig. 6.** Average percentage rates of space control caused by passes of team B.

Examples:

Assume B is attacking with formation 5 against formation 6 of defending A. B is playing a pass without changing its formation. The expected resulting space control then is 18.6%. If B would change its formation to 1 or 3 the results were significantly worse with 10.2% and 11.4%.

If, however, B would switch to formation 3 against formation 6 of A, as it did once, it could improve the resulting space control to 41.7%.

This means that the resulting control rate of a pass can be taken as its success indicator and therefore can be taken for tactical planning and/or simulation: Tactical simulation would take those suggestions in order to optimize frequent offence- or defence-processes or to find out which rare and unexpected activities could generate success by surprising the opponent team (Memmert & Perl, 2009).

## 5    Conclusion

Reducing a complex game to its main patterns and processes can help to analyse its interactions and so improve the understanding of its dynamics and tactical concepts.

Of course, the complex information of 22 players during 2700 seconds per half-time has to be reduced significantly, and therefore the mapping from the real match to the analysis model cannot be correct at every given instance. For example: Mapping the position patterns of tactical groups from 2700 individual situations to only about 20 formation types cannot produce absolute precision. But in turn, much better than the sequence of 2700 situations, it makes visible what the main tactical dynamics is and how successful or unsuccessful processes are.

## References

1. Fonseca, S., Milho, J., Travassos, B., & Araujo, D. (2012). Spatial dynamics of team sports exposed by Voronoi diagrams. *Human Movement Science, 31* (6), 1652-1659.
2. Grunz, A., Memmert, D. & Perl, J. (2012). Tactical pattern recognition in soccer games by means of special self-organizing maps. *Human Movement Science, 31*, 334-343.
3. Kim, S. (2004). Vornoi analysis of a soccer game. *Nonlinear Analysis: Modelling and Control, 9* (3), 233-240.
4. Lees, A., Barton, B. and Kerschaw, L. (2003). The use of Kohonen neural network analysis to establish characteristics of technique in soccer kicking. *Journal of Sports Sciences, 21*, 243-244.
5. Leser, R. (2006). Prozessanalyse im Fußball mittels Neuronaler Netze [Net-based process analysis in soccer], *Human Performance and Sport, 2*, 199-202.
6. Memmert, D. & Perl, J. (2009). Game Creativity Analysis by Means of Neural Networks. *Journal of Sport Science, 27*, 139–149.
7. Perl, J. & Memmert, D. (2011). Net-Based Game Analysis by Means of the Software Tool SOCCER. *International Journal of Computer Science in Sport, 10*, 77-84.
8. Perl, J., Grunz, A. & Memmert, D. (2013). Tactics Analysis in Soccer – An Advanced Approach. *International Journal of Computer Science in Sport, 12*, 1, 33-44.
9. Perl, J., Tilp, M., Baca, A. & Memmert, D. (2013). Neural networks for analysing sports games. T. McGary, P. O'Donoghue and J. Sampaio (eds), *Routledge Handbook of Sports Performance Analysis* (237-247). Abingdon: Routledge.

# An Interval Type-2 Fuzzy Logic Based Classification Model for Testing Single-Leg Balance Performance of Athletes after Knee Surgery

Owais Ahmed Malik and S.M.N. Arosha Senanayake

Faculty of Science, Universiti Brunei Darussalam, Gadong, Brunei Darussalam
{11H1202, arosha.senanayake}@ubd.edu.bn

**Abstract.** Single-leg balance test is one of the most common assessment methods in order to evaluate the athletes' ability to perform certain sports actions efficiently, quickly and safely. The balance and postural control of an athlete is usually affected after a lower limb injury. This study proposes an interval type-2 fuzzy logic (FL) based automated classification model for single-leg balance assessment of subjects after knee surgery. The system uses the integrated kinematics and electromyography (EMG) data from the weight-bearing leg during the balance test in order to classify the performance of a subject. The data are recorded through wearable wireless motion and EMG sensors. The parameters for the membership functions of input and output features are determined using the data recorded from a group of athletes (healthy/having knee surgery) and the recommendations from physiotherapists and physiatrists, respectively. Four types of fuzzy logic systems namely type-1 non-singleton interval type-2 (NSFLS type-2), singleton type-2 (SFLS type-2), non-singleton type-1 (NSFLS type-1) and singleton type-1 (SFLS type-1) were designed and their performances were compared. The overall classification accuracy results show that the interval type-2 FL system outperforms the type-1 FL system in classifying the balance test performance of the subjects. This pilot study suggests that a fuzzy logic based automated model can be developed in order to facilitate the physiotherapists and physiatrists in determining the impairments in the balance control of the athletes after knee surgery.

**Keywords:** single-leg balance, fuzzy logic, classification, knee injury, electromyography, kinematics

## 1    Introduction

Balance control and postural stability are very important for the athletes during sports activities requiring fast movements and changing directions quickly. A complex interaction of central nervous system and musculoskeletal system helps the athletes in maintaining balance during demanding sports actions. A lower limb injury (e.g. anterior cruciate ligament rupture) to an athlete may cause various complications including dynamic joint instability and neuromuscular/proprioception impairments which eventually affect his/her balance control during high-level sporting activities [1, 2]. A

© Springer International Publishing Switzerland 2016
P. Chung et al. (eds.), *Proceedings of the 10th International Symposium on Computer Science in Sports (ISCSS)*, Advances in Intelligent Systems and Computing 392, DOI 10.1007/978-3-319-24560-7_11

85

variety of tests are available to detect the balance/postural control changes after a knee injury or surgery. The single leg balance test measures postural stability (i.e. balance) of a subject and it is a simple, easy and effective method to screen for balance impairments in subjects after knee surgery. The single leg balance testing can be performed with opened or closed eyes and on flat or perturbed surface with different flexion positions of weight-bearing and contralateral knees.

Anterior cruciate ligament (ACL) trauma, being one of the most common knee injuries in sports, alters the proprioceptive function of the joint and leads to changes in joint stability and thus affects the balance control. Proprioceptive structures within knee joint also influence the muscle activity around the joint. In single leg stance, the muscular activity is more intensive as compared to double leg stance due to less redundancy in exoskeleton system. For single leg eyes open and eyes closed balance tests of healthy subjects, a positive correlations between the extent of body sway and the magnitude of muscular activity for tibialis anterior, medial gastrocnemius, quadriceps femoris and gluteus medius muscles has been noted [3, 4]. For ACL injured/reconstructed subjects, the knee stabilizing muscles (vastus medialis, vastus lateralis, semitendinosus, biceps femoris and gastrocnemius) have greater role in single leg balancing task. The amount of activation of these muscles varies with the perturbing stimuli.

In recent past, the physiotherapists and physiatrists have started using the computerized and sensors' based methods for the objective assessment of balance control and postural sway of subjects [5]. Wearable motion sensors (e.g. accelerometers, gyroscopes etc.) have enabled objective measurement of balance control during various clinical tests [5]. However, the use of machine learning techniques with integration of different bio-signals has not been much explored in the previous literature for providing an automated classification of single-leg balance testing performance. In this study, an interval type-2 fuzzy logic (FL) has been used for assessment of single-leg balance performance of subjects after knee surgery (ACL reconstruction). Fuzzy logic is a form of multi-valued logic that helps in approximate reasoning by handling impreciseness and uncertainty present in both quantitative and qualitative information using a single model. FL can handle the impreciseness and uncertainties present in the actual measurements recorded through sensors due to noise and motion artifacts. Type-1 FL systems handle the imprecise data by choosing precise membership function value but these systems lack modeling of uncertainty involved in the measurements, definitions of input and output fuzzy sets, and rules. Thus, a type-2 FL based automated adaptive classification system has been investigated for distinguishing the dynamic balance control of healthy and knee injured subjects at different stages of recovery. The kinematics and EMG signals are recorded through wearable wireless sensors and relevant features are extracted from the data. The parameters for the membership functions (MFs) of antecedents are extracted from the data recorded for healthy and knee injured subjects after surgery during experiments and the parameters for MFs of consequent are determined based on the recommendations/observation of physiotherapists and physiatrists. Theses parameters are tuned by using steepest descent method. The classification performance of the proposed system was tested for singleton and non-singleton inputs and results were compared with type-1 FL system.

The purpose of the system is to provide a complementary decision supporting tool to enable the trainers and physiotherapists to objectively monitor the balance control performance of athletes after knee surgery.

## 2 Methodology

### 2.1 General Framework

A fuzzy logic based general framework for classification of single-leg balance testing performance of subjects after knee surgery is shown in Fig. 1. The framework mainly consists of two modules: data collection and processing of different types of input signals, and an adaptive fuzzy logic based classification module. Although, there is variety of single-leg balance testing activities, but in this study kinematics and neuromuscular features were extracted for eyes-open single-leg balancing on BOSU balance trainer (shaded in Fig. 1) to develop the fuzzy logic based classification model. The output of the fuzzy logic system (FLS) was a defuzzified value transformed into classification (normal/healthy, average or poor) of balance testing performance as provided by two physiotherapists/physiatrists for each trial (see section 2.4).

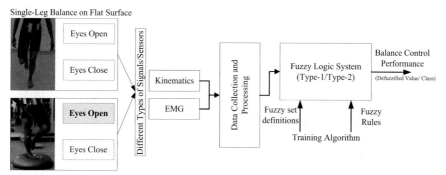

**Fig. 1.** General framework for classification of single-leg balance control of subjects after knee surgery using adaptive fuzzy logic system

### 2.2 Data Collection and Processing

A total 10 subjects (three healthy and seven unilateral ACL reconstructed) were recruited for this study. The wireless micro-electro-mechanical system (MEMS) motion sensors (containing tri-axial accelerometer and gyroscope) from KinetiSense (ClevMed. Inc) were used in this study to collect the kinematics data. A BioCapture physiological monitoring system, consisting of BioRadio and USB receiver, was used to record the EMG signals from the relevant muscles around the knee joint. The motion sensors and EMG electrodes were setup on weight bearing leg of each subject (operated leg for the ACL reconstructed subjects and randomly selected leg for the healthy subjects) as follows: two motion sensors attached to identified positions on lateral aspects of his/her thigh and shank using an adhesive medical tape to note the 3-D kinematics of lower limb extremities, the EMG signals were recorded by placing disposable pre-gelled snap electrodes on vastus lateralis (VL), biceps femoris (BF) and gastrocnemius medialis (GM). The standard guidelines were followed for skin preparation, placements of sensors and electrodes on identified positions and filtering [6]. During each testing session, three trials were performed by every test subject for a duration of at-least 15-20 seconds while each subject was standing on single-leg with eyes open on BOSU balance trainer with knee fully extended for weight bearing leg. The data for multiple trials were collected in order to take care of variability in signals and generating a large dataset containing knee dynamics and neuromuscular signals. A custom software was developed in MATLAB 7.0 for processing of the recorded data. Five features (two kinematics and three EMG), namely mean knee abduction/adduction, range of knee abduction/adduction and normalized root mean squared (RMS) value for VL, BF and GM, were extracted from the processed data for each trial of the subjects as input to design the FLS..

## 2.3    Adaptive Fuzzy Logic Classification System

An adaptive FLS for classification of balance testing performance was designed based on following steps [7].

**Initialize the System.** In order to initialize the type-2 FLS, definitions (types of MFs and their parameters) were determined for antecedents, consequents and inputs. The proposed FLS consisted of five antecedents (mean abduction/adduction, range of abduction/adduction and normalized RMS values of VL, BF and GM muscles) and one output (class for balance testing). The numerical data collected during experiments were used to obtain definitions of antecedents, while the consequent were determined based on the recommendations from physiotherapists and physiatrists for corresponding input. In this context the antecedents and consequent were considered to be type-2 Gaussian with uncertain mean and the input membership functions was type-1 Gaussian for non-singleton inputs. After defining the types of the membership functions, the antecedents' intervals were divided into suitable number of fuzzy sets. In this study, three fuzzy sets (low, average and high) were assumed for all features. The initialization of the parameters (mean, standard deviation etc.) of antecedents' membership functions was done using the procedure described in [7]. The MFs for consequent were defined based on the judgments of physiotherapists and physiatrists about the balance testing for each trial of each subject. A scale of a range 1 through 10 was used to take their input and later mapped to three MFs as *Poor*, *Average* and *Normal/Healthy* to represent the current status of the balance control for each subject. Two types of input membership functions were initialized. In the case of singleton inputs the mean and standard deviation values of each input membership function was the corresponding mean of the 3 trials of the balance testing. For non-singleton inputs, the various measures were represented by type-1 Gaussian MFs. The mean and standard deviation value of each input membership function were the average of means and standard deviation of various measures for a particular subject, respectively.

**Create Rule Base.** Initially a total of 243 (3×3×3×3×3) rules were generated based on all possible combinations of five antecedents and then irrelevant rules were eliminated with a semi-automatic rule-pruning mechanism. The consequent of each rule was decided based on the feedback from physiotherapists and physiatrists.

**Train Rules and Validate Trained Rules.** The rules were trained in order to improve their accuracy in classifying the balance testing performance. The parameters of various membership functions were modified by propagating the inputs through FLS based on computed error and steepest descent approach [8]. The data recorded from subjects were used as training dataset. This dataset contained the input-output pair where the inputs were kinematics and EMG measurements for each subject and the output was the evaluation (1-10) of balance testing performance provided by the physiotherapists and physiatrist. The validation of the system was done by using the Leave-One-Out Cross Validation (LOOCV) method. This method was suitable for the small sample size used in this study. In LOOCV method, the FLS was trained on N-1 samples from the dataset as described above and one sample was left as the validation sample. This process was repeated N times and the overall classification accuracies and error measurements were computed for fuzzy logic systems based on singleton and non-singleton inputs.

**Table 1.** Percentage of Classification Accuracy of Different Types of Subjects Using Height Reduction and Modified Height Reduction Defuzzification Methods

| Type of FLS | Subjects | Height Reduction | Modified Height Reduction |
|---|---|---|---|
| NSFLS Type-2 | Healthy | 84.76 | 84.72 |
| | Average | 76.93 | 77.06 |
| | Poor | 75.85 | 76.14 |
| SFLS Type-2 | Healthy | 83.76 | 83.72 |
| | Average | 76.06 | 76.37 |
| | Poor | 73.54 | 73.64 |

# 3 Results

The system was designed and tested using the data collected from the healthy and ALC reconstructed subjects. The healthy subjects were having a mean age of 27.00± 2.15 years, mean height 163.20±8.45 cm, and mean weight 62.21±10.79 kg. The ACL reconstructed subjects were at different stages of rehabilitation (from 2 months to around 1 year after ACL reconstruction) with mean age: 29.45± 3.63 years, mean height 162.70±8.25 cm, and mean weight 69.30±10.25 kg. Four types of fuzzy logic systems namely type-1 non-singleton interval type-2 (NSFLS type-2), singleton type-

2 (SFLS type-2), non-singleton type-1 (NSFLS type-1) and singleton type-1 (SFLS type-1) were designed and their performances were compared. Mamadani type inference system was used for all FL systems with product implication and t-norm fuzzy operations. Further, the classification performance of two different type-reduction/defuzzification methods (height reduction and modified height reduction) was tested for interval type-2 FL systems. For type-1 FL systems, the height method was used for the defuzzification process. Each FLS was trained and tested (using LOOCV method) with a dataset of size 10×3. Fig.2 shows a comparison of overall classification accuracy based on defuzzified values for balance testing of healthy and ACL reconstructed subjects using type-1 and type-2 (using height reduction method) FL systems before and after training. An increase in the classification accuracy has been found for all FL systems except for SFLS Type-1. The classification accuracy values for three (NSFLS type-2, SFLS type-2 and NSFLS type-1) FL systems were found higher (with maximum value of approximately 80% for NSFLS type-2) as compared to SFLS type-1 system because these three types of FL systems can handle the stationary noise in measurements and noise in training/testing data. Fig. 2 depicts that both type-2 systems have performed better than type-1 systems which indicates that type-2 FLS handles uncertainty better than type-1 FLS and the outputs produced by type-2 FLS were much closer to the actual outputs. A comparison of height defuzzification and modified height defuzzification method was also performed for typ-2 FL systems. Fig. 3 depicts that there is no significant difference in terms of performance of modified height defuzzification and height defuzzification methods while classifying the balance testing of healthy and ACL reconstructed subjects. The average of mean square error (MSE) is depicted in Fig. 4 for all FL systems. Based on the recorded data, three groups of subjects were identified and the classification accuracy of each group for different types of FL systems is shown in Table 1. The accuracy of the system was comparable with the previously reported results using ANFIS model [6] while using less number of features.

## 4    Conclusion

This study investigated the design and application of an automated interval type-2 fuzzy logic system for modeling the uncertainty in the balance testing data and classifying the postural control of healthy and knee injured subjects after surgery. Type-2 fuzzy logic systems performed superior as compared to type-1 fuzzy logic systems in detecting the balance control of the subjects based on the selected kinematics and EMG input parameters. These findings suggest that the uncertainty present in measurements and observations can be better handled using the type-2 FL as compared to using the type-1 FL. Moreover, this study shows that an FL based automated decision support system is feasible in assessing the balance control of the athletes using kinematics and EMG data. Such a system can be useful for physiatrists, physiotherapists and sports trainers in addition to relying on their observations and other balance testing tools. However, the maximum overall classification accuracy of approximately 80% was achieved which requires further investigations in choosing the appropriate input features and designing of rules. Further, the system will be enhanced and tested for examining the balance control of athletes on various balance testing platforms.

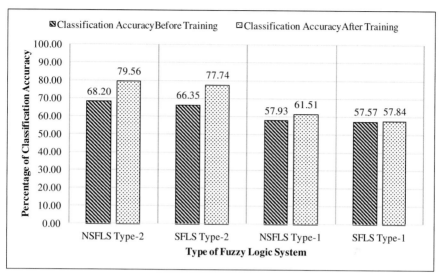

**Fig. 2.** Comparison of percentage of classification accuracy before and after training for all FL systems

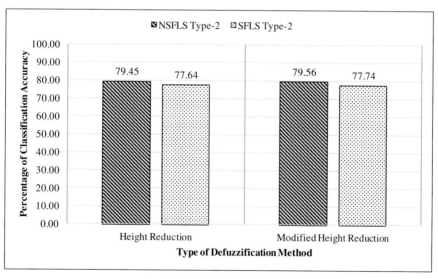

**Fig. 3.** Comparison of percentages of overall classification accuracy using different defuzzification/type reduction methods

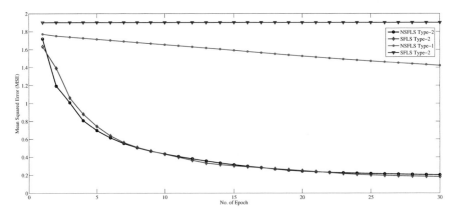

**Fig. 4.** Average mean square error (MSE) for type-1 and type-2 FL systems for during training phase

# References.

1. Thomas AC, Villwock M, Wojtys EM et al (2013) Lower extremity muscle strength after anterior cruciate ligament injury and reconstruction. *J Athl Train*, 48:610-620
2. Paterno MV, Schmitt LC, Ford KR et al (2010) Biomechanical measures during landing and postural stability predict second anterior cruciate ligament injury after anterior cruciate ligament reconstruction and return to sport. *Am J Sports Med*, 38:1968-1978
3. Suponitsky Y, Verbitsky O, Peled E et al (2008) Effect of selective fatiguing of the shank muscles on single-leg-standing sway. *J Electromyogr Kinesiol*, 18:682-689
4. Levin O, Mizrahi J, Adam D et al (2000) On the correlation between force plate data and emg in various standing conditions. Paper presented at Proceedings of the fifth annual conference of the international functional electrical stimulation society, Center for Sensory-Motor Interaction, Aalborg University, Denmark, 18-20 June 2000
5. Mancini M, Horak FB (2010) The relevance of clinical balance assessment tools to differentiate balance deficits. *Eur J Phys Rehabil Med*, 46:239-248
6. Malik OA, Senanayake SMN, Zaheer D (2014) An intelligent recovery progress evaluation system for acl reconstructed subjects using integrated 3-d kinematics and emg features. *IEEE Journal of Biomedical and Health Informatics*
7. Malik OA, Senanayake SMN, Zaheer D (2014) An adaptive interval type-2 fuzzy logic framework for classification of gait patterns of anterior cruciate ligament reconstructed subjects. Paper presented at IEEE International Conference on Fuzzy Systems (FUZZ-IEEE), Beijing, China, 6-11 July 2014
8. Mendel JM (2001) Uncertain rule-based fuzzy logic systems: Introduction and new directions. Prentice Hall PTR

# A Comparison of Classification Accuracy for Gender Using Neural Networks Multilayer Perceptron (MLP), Radial Basis Function (RBF) Procedures Compared to Discriminant Function Analysis and Logistic Regression Based on Nine Sports Psychological Constructs to Measure Motivations to Participate in Masters Sports Competing at the 2009 World Masters Games

Ian Heazlewood [1], Joe Walsh [1], Mike Climstein [2], Jyrki Kettunen [3], Kent Adams [4] and Mark DeBeliso [5]

[1]Charles Darwin University
[2]University of Sydney
[3]Arcada University of Applied Sciences
[4]California State University
[5]Southern Utah University

**Abstract.** Neural networks can be applied to many predictive data mining applications due to their power, flexibility and relatively easy operations. Predictive neural networks are very useful for applications where the underlying process is complex, such as in classification using a mix of nominal and ratio level variables and for predictive validity based on classification modelling. A neural network can approximate a wide range of statistical models without requiring the researcher to hypothesize in advance certain relationships between the dependent and independent variables. The two major applications are multilayer perceptron (MLP) and radial basis function (RBF) procedures. In contrast to MLP networks, in the RBS networks it is only the output units that have a bias term. Discriminant analysis (or discriminant function analysis) based on classification modelling is applied to classify cases into the values of a categorical dependent variable, usually a dichotomy. Logistic regression is useful for situations in which you want to be able to predict the presence or absence of a characteristic or outcome based on values of a set of predictor variables. It is similar to a linear regression model but is suited to models where the dependent variable is dichotomous. The aim of this research was to apply both neural networks, discriminant function analysis (a more traditional statistical approach under the general linear model) and logistic regression and compare their ability as statistical techniques to classify the different genders based nine sports psychological

© Springer International Publishing Switzerland 2016
P. Chung et al. (eds.), *Proceedings of the 10th International Symposium on Computer Science in Sports (ISCSS)*, Advances in Intelligent Systems and Computing 392, DOI 10.1007/978-3-319-24560-7_12

constructs to measure motivations to participate in masters sports. The sample consisted of 3687 male and 3488 female master's athletes who participated in the 2009 World Masters Games and represented a volunteer/convenient sample in the study and a cross-sectional non-experimental research design. The Motivations of Marathoners Scales (MOMS) psychometric instrument assessed participant motivation by nine constructs/factors using factor scores from a 56 item seven Likert type survey instrument measuring motivations to participate. These factors were health orientation, weight concern, personal goal achievement, competition, recognition, affiliation, psychological coping, life meaning and self-esteem. The accuracy of the solutions were assessed with neural networks, by classification accuracy using both test and holdout samples, predicted-by-observed chart, ROC curve, cumulative gains and lift charts, independent variable importance and normalised importance; and discriminant function analysis by both original and cross-validation samples, lambda values, p-values, tolerance, F to remove and in stepwise discriminant analysis by the hierarchy of inclusion steps. Similar methods were applied when assessing classification accuracy using logistic regression. The results in terms of MLP analysis was overall correct percent of 64.4% and the order of importance was competition, self-esteem, affiliation, recognition, weight concern, health orientation, goal achievement, psychological coping and life meaning. In terms of RBF analysis training sample overall correct percent was 60.5% and order of importance was competition, affiliation, recognition, psychological coping, weight concern, life, meaning, self-esteem, goal achievemnt and health orientation. For the discriminant analysis the overall correct classification rate was 63.3% and for logistic regression 63.0% and the stepwise entry order into both analyses was affiliation, competition, self-esteem, recognition, weight concern and health orientation. The classification accuracies based on MLP, discriminant analysis and logistic regression were very similar in outcome for both the classification of gender and combined classification accuracy. None of the classification techniques based on neural network analyses and multivariate method of discriminant analysis and logistic regression were overtly superior to each other. Although it is important to note the RBF neural network displayed classification accuracy slightly lower than the other three methods.

# 1 Introduction

## 1.1 Neural Networks

Neural networks can be applied to many predictive data mining applications due to their power, flexibility and relatively easy operations. Predictive neural networks [1, 2] are very useful for applications where the underlying process is complex, such as in classification using a mix of nominal and ratio level variables and for predictive validity based on classification modelling. A neural network can approximate a wide range of statistical models without requiring the researcher to hypothesize in advance certain relationships between the dependent and independent variables. Neural networks are the preferred tool for many predictive data mining applications because of their power, flexibility, relevance and ease of use. Predictive neural networks are particularly useful in applications where the underlying process is complex, especially pattern recognition and classification problems that are based on predictive and concurrent validity.

Neural networks used in predictive applications, such as the multilayer perceptron (MLP) and radial basis function (RBF) networks, are supervised in the sense that the model-predicted results can be compared against known values of the target variables. These target variables are identified on a priori criteria by the researcher. The term neural network applies to a loosely related family of models, characterized by a large parameter space and flexible structure, descending from studies of brain functioning. As the family grew, most of the new models were designed for non-biological applications, though much of the associated terminology reflects its origin in biology [1, 2]. A neural network is a massively parallel distributed processor that has a natural propensity for storing experiential knowledge and making it available for use and is analogous to human brain function. Specifically, it resembles the brain in two respects; knowledge is acquired by the network through a learning process and Interneuron connection' strengths, known as synaptic weights and analogous to human synapses, are used to store the knowledge.

A neural network can approximate a wide range of statistical models without requiring that you hypothesize in advance certain relationships between the dependent and independent variables, a non *a priori* model. Instead the form of the relationships is determined during the learning process [1, 3]. A type of neural processing phenomenology in this context. The trade-off for this flexibility is that the synaptic weights of a neural network are not easily interpretable. Thus, if you are trying to explain an underlying process that produces the relationships between the dependent and independent variables, it would be better to use a more traditional statistical model, such as discriminant analysis or logistic regression. However, if model interpretability is not important, you can often obtain good model results more quickly using a neural network [1, 3]. Although neural networks impose minimal demands on model structure and assumptions, unlike inferential statistics, it is useful to understand the general neural architecture or neural network structure. The multilayer perceptron (MLP) and radial basis function (RBF) networks are functions of predictors (also called inputs or independent variables) that minimize the prediction error of target variables (also called outputs) [1, 3].

## 1.2    Discriminant Analysis

Discriminant analysis (or discriminant function analysis) based on classification modelling is applied to classify cases into the values of a categorical dependent variable, usually a dichotomy [3. 4]. In sport this could be males compared to females on different motor fitness tests or different player grades using the same principles. If discriminant function analysis is effective for a set of data, the classification table of correct and incorrect estimates will yield a high percentage correctly classified cases and maybe useful in such processes as sport talent identification, such as in Olympic sports or motor fitness differences based on gender. The major foci of discriminant analysis [3, 4, 5, 6, 7] are to; 1. Classify cases into groups using a discriminant prediction equation and test theory by observing whether cases are classified correctly as predicted. Investigate differences between or among groups and determine the most parsimonious way to distinguish among groups. 2. Determine the percent of variance in the dependent variable explained by the independents. Determine the percent of variance in the dependent variable explained by the independents over and above the variance accounted for by control variables, using sequential discriminant analysis. 3. Assess the relative importance of the independent variables in classifying the dependent variable and discard variables, which are little related to group distinctions.

## 1.3    Logistic Regression

Logistic regression is useful for situations in which you want to be able to predict the presence or absence of a characteristic or outcome based on values of a set of predictor variables. It is similar to a linear regression model but is suited to models where the dependent variable is dichotomous. Logistic regression coefficients can be used to estimate odds ratios for each of the independent variables in the model. Logistic regression is applicable to a broader range of research situations than discriminant analysis. When comparing logistic regression with discriminant analysis both provide similar predictive and classification outcomes and employ similar diagnostic measures such as model fit indices, however logistic regression is less affected when basic assumptions as deviations from normality occur [8].

Previous research evaluated classification accuracy using MLP and discriminant analysis to classify high ability and non-high ability karate athletes based on differences on physiological and biomechanical measures [9], classification accuracy of MLP, RBF and discriminant analysis using similar predictors of participant motivation when classifying gender based on this construct [10] and a comparison of classification accuracy for gender using MLP and RBF and discriminant analysis based on biomechanical measures of isokinetic torque, work, power, fatigue index, counter movement jump and 10m acceleration [11]. These research studies discovered the classification accuracy using MLP, RBF and discriminant analysis produced very similar results, however these studies used smaller sample sizes and did not include the method of logistic regression as a multivariate statistical classification method. Using and comparing multiple classifications methods as MLP, RBF, discriminant analysis and logistic regression would enable a more complete understanding of the value of such methods when attempting to classify into mutually exclusive groups based on underpinning differences, if they exist, between different groups.

## 1.4    Research Aim

The aim of this research was to apply both neural networks MLP and RBF and standard statistical multivariate discriminant function analysis (a more traditional statistical approach under the general linear model) and logistic regression, and then compare their ability as statistical techniques to classify the different genders based nine sports psychological constructs/factors to measure motivations to participate in masters sports. Specifically, the motivations to participate factors were health orientation, weight concern, personal goal achievement, competition, recognition, affiliation, psychological coping, life meaning and self-esteem [12]. In addition, the research aim also focused on which factors or variables provided the greatest difference between the genders, the establishment of a hierarchy of factor-variable importance and to assess which multivariate method of classification provided the best solution.

## 2    Methods

— The sample consisted of 3687 male (age = 53.72 years, s.d. = +/-10.05 years) and 3488 female (age = 49.39 years, s.d. = +/-9.15 years) master's athletes and represented a volunteer/convenient sample in the study and a cross-sectional non-experimental research design from the potential population of  approximately 33,000 masters athletes competing at the 2009 World Masters Games.

MLP and RBF neural networks, discriminant analysis and logistic regression were applied, based on a set of dependent/covariate variables, which were the nine participant motivations factors of health orientation, weight concern, personal goal achievement, competition, recognition, affiliation, psychological coping, life meaning and self-esteem. The categorical or classification dichotomous variable utilised in the different models was gender, specifically the male and female athletes who competed at the 2009 World Masters Games and who completed the sports psychometric instrument specifically designed to measure the nine different participant motivation factors [9]. The instrument assessed the participant motivation factors derived using factor scores from a 56 item question bank and athletes responding on a seven point Likert type scale for each question [9]. The sport psychological instrument was completed via an online survey using the Limesurvey[tm] interactive survey system prior to, and during competition at the 2009 World Masters Games. The four statistical methods were compared for their classification accuracy to successfully discriminate between male and female athlete responses. Neural networks, specifically the multilayer perceptron (MLP) and radial basis function (RBF) networks, were applied and compared with stepwise method discriminant analysis and stepwise logistic regression for classification accuracy.  The neural network multilayer perceptron architecture was based on:

- Selecting one hidden layer where the hidden layer contains unobservable network nodes (units). Each hidden unit is a function of the weighted sum of the inputs. The function is the activation function, and the values of the weights are determined by the estimation algorithm.
- The selected activation function was the hyperbolic tangent, where the activation function links the weighted sums of units in a layer to the values of units in the succeeding layer.
- Hyperbolic tangent function has the form $\gamma(c) = \tanh(c) = (e^c - e^{-c}) / (e^c + e^{-c})$. (1)

- It takes real-valued arguments and transforms them to the range $(-1, 1)$. When automatic architecture selection is used in SPSS, this is the activation function for all units in the hidden layers.

The identity function was selected and this function has the form: $\gamma(c) = c$. It takes real-valued arguments and returns them unchanged. When automatic architecture selection is used, this is the selected activation function for units in the output layer if there are any scale-dependent variables. Training the network was based on the batch method. This method updates the synaptic weights only after passing all training data records, which means batch training uses information from all records in the training dataset. Batch training is often preferred because it directly minimizes the total error and is most useful for smaller datasets. In contrast to MLP networks, in the RBF networks it is only the output units that have a bias term.

Discriminant analysis was based on using the nine factors/variables as the starting point in the stepwise method, which is based on statistical criteria to enter the model at each calculation step. Gender was used as the independent dichotomous variable in the analysis. It must be emphasised that the comparison of discriminant analysis with neural network analysis were based on the identical nine factors. The stepwise method was applied to generate a hierarchy of importance in terms of predictor variables and to assess which variables contributed to significant difference between genders.

Applying logistic regression the dependent variable was the nominal dichotomous variable gender where dummy coding zero for male and one for female. The predictor or independent variables were the nine factors of participant motivation utilised in the MLP, RBF and discriminant analysis. The logistic regression variable selection method was the forward selection likelihood ratio stepwise selection method with entry testing based on the significance of the score statistic, and removal testing based on the probability of a likelihood-ratio statistic based on the maximum partial likelihood estimates.

## 3    Results

The different statistical methods produced slightly different outcomes in independent variable importance and slightly different classification accuracy. For MLP the order of importance competition, self-esteem, affiliation, recognition, weight concern, health orientation, goal achievement, psychological coping and life meaning. For RBF the order of importance is somewhat different and the order was competition, affiliation, recognition, psychological coping, weight concern, life, meaning, self-esteem, goal achievemnt and health orientation. Self-esteem which was identified as number two with MLP is identified as number seven with RBF, whereas life meaning is ranked ninth with MLP and ranked sixth with RBF. This indicates that the two approaches are providing different solution in terms of the predictor variable importance hierarchy.

The discriminant and logistic regression analyses in terms of the factors entered in the stepwise solutions were identical and the factors were entered in the following order of affiliation, competition, self-esteem, recognition, weight concern and health orientation.

The discriminant and logistic regression analyses in terms of the factors entered in the stepwise solutions were identical and the factors were entered in the following order of affiliation, competition, self-esteem, recognition, weight concern and health orientation. It is important to highlight that the stepwise methods only selected six significant factors as important to providing the classification solutions, whereas the MLP and RBF methods included all nine factors in the analysis with no inclusion or exclusion criteria that exist with discriminant analysis and logistic regression.

In terms of classification accuracy MLP, discriminant analysis and logistic regression were very similar at approximately 66% for males and approximately 60% for females and the exact classification percentages are displayed in table 1. The RBF classification accuracy was slightly less accurate at 63.5% for males and 57.4% for females. The combined classification accuracy was MLP 64.4%, RBF 60.5%, discriminant analysis 63.3% and logistic regression 63.0%, and once again the RBF function was slightly less accurate for females. Overall the actual differences when comparing the accuracy of the different classification methods were only marginal in outcome. This indicates a convergence in all classification solutions with the data set in this research.

Table 1. Classification accuracy for males, females and combined genders based on MLP, RBF, discriminant analysis and logistic regression.

| Method | Male Classification % | Female Classification % | Combined Classification % |
|---|---|---|---|
| MLP | 66.8 | 61.8 | 64.4 |
| RBF | 63.5 | 57.4 | 60.5 |
| Discriminant | 66.0 | 60.5 | 63.3 |
| Logistic | 66.1 | 59.8 | 63.0 |

## 4    Conclusion

The multilayer perceptron (MLP) networks, was more effective in predicting group membership based on gender using the nine factors that represented the multiple dimensions of participant motivation within male and females athletes competing at the 2009 World Masters Games (WMG) and displayed a reasonable level of predictive validity and marginally more predictive than radial basis function (RBF) networks. However, when MLP is compared to the general linear multivariate methods of discriminant analysis and logistic regression MLP only marginally outperforms these methods. The MLP and RBF utilised the nine factors in the analysis whereas stepwise discriminant analysis and logistic regression required only six discriminating variable to provide a solution nearly as accurate as MLP.

In terms of which participant motivation factors were the best discriminators between the genders, both discriminant analysis and logistic regression produced identical hierarchies concerning order of importance of factors, which were in order from the most to least important affiliation, competition, self-esteem, recognition, weight concern and health orientation. The variables excluded from the model as not contributing significantly were life meaning, psychological coping and goal achievement. The order of importance identified by MLP and RBF were different from discriminant analysis and logistic regression, as two slightly difference orders of importance were derived.

The order of importance MLP was completion, self-esteem, affiliation, recognition, weight concern, health orientation goal achievement, psychological coping and life meaning, whereas for RBF the order was competition, affiliation, recognition, psychological coping, weight concern, life meaning, self-esteem, goal achievement and health orientation. This indicates although the two methods were very different concerning order of importance they were somewhat similar in classification accuracy. One of the problems with neural networks is they can produce different solutions from the same data base as they open ended learning structures and hence the possibility of multiple solutions based on the identical data base. To overcome this problem in terms of replicating results the researcher has to use the same initialization value for the random number generator, the same data order, and the same variable order, in addition to using the same procedure settings [3]. Alternatively, as stated in the introduction if the researcher is trying to explain an underlying process that produces the relationships between the dependent and independent variables, it would be better to use a more traditional statistical model, such as discriminant analysis or logistic regression and in this research these methods produced essentially identical solutions.

# References

1. Fausett, L.: Fundamentals of Neural Networks: Architectures, Algorithms and Applications. Upper Saddle River NJ: Prentice Hall, (1994)
2. SPSS Inc.: SPSS Statistics Base User's Guide 17.0. Users Guide. Chicago, IL: SPSS Inc, (2007)
3. SPSS Inc.: SPSS Neural Networks™ 17.0. Chicago, IL: SPSS Inc, (2007)
4. Norusis, M.: Advanced Statistics Guide: SPSSX. Chicago, IL: SPSS Inc, (1985)
5. StatSoft, Inc.: Electronic Statistics Textbook. Tulsa, OK: StatSoft. WEB: http://www.statsoft.com/textbook/, (2010)
6. Hair, J., Block, W., Babin, B., Anderson, R., Tatham, R.: Multivariate Data Analysis. (6th Ed.). Upper Saddle River: Pearson - Prentice Hall,( 2006)
7. SPSS Inc.: IBM SPSS Statistics 22. Chicago, IL: SPSS Inc, (2013)
8. SPSS Inc.: SPSS Regression 17.0. Chicago, IL: SPSS Inc, 2007.
9. Heazlewood, I., Keshishian, H.: A Comparison of Classification Accuracy for Karate Ability Using Neural Networks and Discriminant Function Analysis Based on Physiological and Biomechanical Measures Of Karate Athletes. Refereed Proceedings of the Tenth Australasian Conference on Mathematics and Computers in Sport. July 5-7, 2010. Crowne Plaza, Darwin, Northern Territory, Australia. Pp. 197-204. (2010)

10. Heazlewood, I., Walsh, J., Burke, S., Climstein, M., Kettunen, J., Adams, K., DeBeliso, M.: A Comparison of Classification Accuracy for Gender Using Neural Networks Multi-layer Perceptron (MLP) and Radial Basis Function (RBF) Procedures and Discriminant Function Analysis Based On Nine Sports Psychological Constructs to Measure Motivations to Participate in Masters Sports. Proceedings of 2012 Pre-Olympic Congress-IACSS 2012, pp. 88-94: Liverpool, England, UK, July 24-25, 2012 ISBN 978-1-84626-094-0. (2012)

11. Heazlewood, I., Walsh, J.: A Comparison of Classification Accuracy for Using Neural Networks Multilayer Perceptron (MLP) and Radial Basis Function (RBF) Procedures and Discriminant Function Analysis. Proceedings of the International Association of Computer Science in Sport Conference (IACSS2014). Ed. Assoc. Prof. Ian Heazlewood, Assoc. Prof. Anthony Bedford, Darwin, Australia, June 22-24. 2014. Pp. 116-120. (2014)

12. Masters, K., Ogles, B., Dolton, J.: The development of an instrument to measure motivation for marathon running: the Motivations of Marathoners Scales (MOMS). Research Quarterly in Exercise and Sport. 1993, 64 (2):134-43. (1993)

# Detection of Individual Ball Possession in Soccer

Martin Hoernig[1] (HOERNIG@IN.TUM.DE),
Daniel Link[2] (DANIEL.LINK@TUM.DE),
Michael Herrmann[1](MICHAEL.HERRMANN@TUM.DE),
Bernd Radig[1](RADIG@IN.TUM.DE),
Martin Lames[2](MARTIN.LAMES@TUM.DE)

[1] Image Understanding and Knowledge-Based Systems, TUM,
[2] Department of Performance Analysis, TUM,
Arcisstraße 21, D-80333 München, Germany

**Abstract.** While ball possession usually is considered on team level, a model on player level brings several advantages. We calculate ball possession and control statistics for all players as well as new ball control heat maps to evaluate the players' performances. Furthermore, a basis for detecting events and tactical structure becomes available. To derive individual ball possession from spatio-temporal data, we present an automatic approach, based both on physical knowledge and machine learning techniques. Moreover, we introduce different ball possession definitions and algorithms to model various grades of ball control. When applied to flawless raw data, the algorithms show precision and recall ratios between 80 and 92 %. With approximately four percentage points less in uncorrected data, the presented algorithms are also reliable in real-world scenarios.

## 1 Introduction

Game analysis plays an important role in soccer coaching. Observing and analyzing tactical behavior can generate useful information that can be used for managing training processes and developing match strategies [1]. The technological innovations of recent years in the field of position tracking present new challenges in analyzing and interpreting the resulting data. The key lies in using intelligent algorithms in order to derive complex performance indicators from the raw data that add real value when it comes to game analysis [4, 5].

This paper describes and evaluates a method that enables different types of ball possession to be detected using ball and player positions. From a sports science perspective, ball possession is the most commonly investigated performance indicator [8]. Its relevance is easy to understand, since being in control of the ball is a fundamental prerequisite for being able to invade the opposing team's third of the pitch and score goals. Existing research [6, 7, 9] has been based exclusively on ball possession on the team level. Because up until now, such data has been collected by the *competition information providers (CIP)* solely on the basis of ball possession changes between teams. The reason for this

© Springer International Publishing Switzerland 2016                                    103
P. Chung et al. (eds.), *Proceedings of the 10th International Symposium on Computer Science in Sports (ISCSS)*, Advances in Intelligent Systems and Computing 392, DOI 10.1007/978-3-319-24560-7_13

reduction in complexity is that ball possession data is collected by human data loggers concurrent with the game, and it would be too involved and expensive to manually record data on an individual player basis.

## 2 Ball Possession - Models and Detection

When it comes to the definition of ball possession, there is a certain amount of flexibility. In the following we use these definitions which describe various grades of ball control:

1. *Individual Ball Possession (IBP)* begins at the moment a player is able to perform an action with the ball following an IBP of another player or a game interruption. It ends at the moment an IBP begins for another player.
2. *Individual Ball Action (IBA)* of a player begins at the moment this player is able to perform an action with the ball and had no prior IBA. It ends at the moment the player is no longer able to perform any further action with the ball or ends with the next game interruption.
3. *Individual Ball Control (IBC)* for a player begins when an IBA for this player begins and ends at the moment this particular IBA ends. In difference to the IBA, an IBC only takes place if the player is able to decide between several strategic courses of action during the IBA.

Accordingly, a successful passing involves an IBP lasting until the ball is received by another player, while IBA and IBC end when the passer is no longer able to interact with the ball. No IBC occurs if the player has only the option of this passing.

Automatic detection of ball possession involves a three-step process. In the first step, the spatio-temporal tracking data provided by CIPs are pre-processed by a Rauch-Tung-Striebel smoother [10, Ch. 3, 4, 8]. The next step is the core of the procedure: the detection of IBP and IBA start and end points, concluded by the final step, an estimation of ball control using a Bayesian network.

By using the IBP model, it is sufficient to only calculate the moments in which the IBPs start, given that the CIP provides a running flag containing a match's status (running or interrupted) for every point in time. Because of a missing z-coordinate in common CIP data, a component of the distance between the players and the ball stays unknown. Thus, a threshold based detection is prone to errors. Instead, we use local maxima of the amplitude of changes in ball velocity (ball accelerations) to detect kicks. If a kick is detected, a player is within a distance that allows a physical interaction, and if they are the player with the shortest distance to the ball, IBP is assigned to the player. We refer to this method as *kick detection*.

Following the definition, the kick detection can also be applied to calculate IBA start points. There are three possible ways an IBA for a player can end: 1) the game is interrupted, 2) another player gains IBA or 3) the player is no longer able to interact with the ball. Whereas the first case is trivial to detect because of the running flag data, and the second case can be obtained directly

Table 1: IBP and IBA detection results in % (0.6 s tolerance window)

|                                           | Recall | Precision |
|-------------------------------------------|--------|-----------|
| **IBP**                                   |        |           |
| Net play time without tracking errors     | 80.1   | 86.1      |
| Net play time                             | 78.0   | 76.9      |
| **IBA**                                   |        |           |
| Net play time without tracking errors     | 86.8   | 92.4      |
| Net play time                             | 88.0   | 86.6      |

through kick detection results, only the last case requires special treatment. This involves checking whether a player will still be able to interact with the ball in the near future. Our approach makes use of the current position and velocity of the ball to give an estimate of its future location. As long as it is possible for the player currently in possession of the ball to control the ball in one second (the moment of prediction), they will retain IBA. The ability to control a ball is again checked via a distance threshold. We refer to this method as *ball prediction*.

Once IBA start and end points are known, the IBA intervals can be derived. We distinguish ball control IBAs from IBAs without courses of strategic interaction by categorizing them into IBCs and non-IBCs. For this reason, we decided to train a Bayesian network [3] to classify ball control based on a set of variables like duration of ball action, average ball velocity and acceleration, variance of ball velocity and acceleration, average distance between ball and the player in ball possession, and number of opposing players within certain distances.

## 3   Evaluation

A match in a top European league served as a test sample in order to evaluate the quality of IBP, IBA, and IBC detection as well as a basis to calibrate the presented thresholds. The ball possession data were manually annotated by a trained, independent observer after the game to form a ground truth. Table 1 shows detection rates for sequences without errors which were caused by the underlying tracking system. These are compared to the results logged throughout the complete net play time. Precision is the ratio of correctly identified possession changes to the total number of changes in the measurements, recall is the ratio between the correctly identified changes to the total number of changes in the ground truth. By way of comparison, the CIP's team ball possession achieved a precision of 52.2 % to the given ground truth.

The IBC recognition rates were also determined by a ground truth comparison. The degree of consistency according to Cohen [2] is $\kappa = 0.38$ along with 92 % accordance to the ground truth. However, with only 25 non-IBC intervals, the used training set may not be sufficient for the proposed machine learning approach.

Fig. 1: Heat maps of a center-forward based on all positions during the game (a) compared to positions during IBC (b). Their team played from right to left.

(a) Movement                                    (b) Ball control

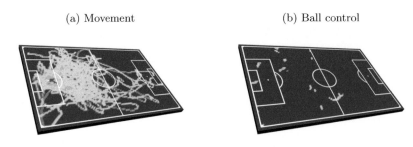

## 4 Applications

IBP can be used to describe the characteristic of performance during a game. The analyzed match had a duration of 90:12 min and a net game duration (excluding stoppages) of 57:56 min. IBC was present in 1,291 phases totaling 29:48 min. The home team had a greater share of IBP, especially in the first half, whereas the away team controlled the ball for only 2:57 min. The average IBC duration, i.e. the time interval in which a player in possession of the ball could make and execute a tactical decision, was 1:27 min for field players who played for the entire match. There are, however, large differences between the players covering from 0:22 up to 3:38 min.

Fig. 1 shows a "traditional" heat map for a center-forward in contrast to a heat map based only on the time periods where the player had IBC. This player had the shortest average IBC intervals (0.9 s) in his team, which is not surprising for a center-forward.

## 5 Conclusion

Nowadays, the quality of tracking allows for individual ball possession to be reliably detected. Using the proposed methods on uncorrected data results in precision and recall ratios of around 80 %. This outcome indicates a wide variety of potential applications. Among statistical analyses and visual representations basic event detectors can be built easily based on individual possession data. Events such as passes, tackles, or shots on goal can be deduced directly. Also, being able to detect ball possession is a fundamental prerequisite for being able to discern higher value tactical structures like availability, pressing strategies, or marking tactics. The capability to recognize ball possession types holds considerable potential for improving the quality of match analysis in professional soccer. Further research has to be done in order for its importance to be assessed. Automation enables an additional performance-relevant parameter to be detected and evaluated in large data samples without the need for additional resources.

# References

1. Carling, C., Reilly, T., Williams, A.M.: Performance assessment for field sports: physiological, and match notational assessment in practice. Routledge (2009)
2. Cohen, J.: A coefficient of agreement for nominal scales. Educational and Psychological Measurement 20, 37–46 (1960)
3. Cooper, G.F., Herskovits, E.: A bayesian method for the induction of probabilistic networks from data. Machine learning 9(4), 309–347 (1992)
4. Grunz, A., Memmert, D., Perl, J.: Tactical pattern recognition in soccer games by means of special self-organizing maps. Human movement science 31(2), 334–343 (2012)
5. Gudmundsson, J., Wolle, T.: Football analysis using spatio-temporal tools. Computers, Environment and Urban Systems 47, 16–27 (2014)
6. Hughes, M., Franks, I.: Analysis of passing sequences, shots and goals in soccer. Journal of Sports Sciences 23(5), 509–514 (2005)
7. Jones, P., James, N., Mellalieu, S.D.: Possession as a performance indicator in soccer. International Journal of Performance Analysis in Sport 4(1), 98–102 (2004)
8. Mackenzie, R., Cushion, C.: Performance analysis in football: A critical review and implications for future research. Journal of sports sciences 31(6), 639–676 (2013)
9. Pratas, J., Volossovitch, A., Ferreira, A.: The effect of situational variables on teams performance in offensive sequences ending in a shot on goal. a case study. Open Sports Sciences Journal 5, 193–199 (2012)
10. Särkkä, S.: Bayesian filtering and smoothing, vol. 3. Cambridge University Press (2013)

# Towards Better Measurability - IMU-Based Feature Extractors For Motion Performance Evaluation

Heike Brock and Yuji Ohgi

Keio University, Graduate School of Media and Governance,
Fujisawa-shi, Kanagawa, Japan

**Abstract.** Capturing human motion performances with inertial measurement units constitutes the future of mobile sports analysis, but requires sophisticated methods to extract relevant information out of the sparse and unintuitive inertial sensor data. Kinematic data like body joint positions and segment orientations can be estimated from a sensor's accelerations and angular velocities. For further analysis, it is necessary to develop intelligent retrieval strategies that can make sense of the underlying motion information. In this paper, we therefore discuss how to retrieve main motion determinants from raw and processed inertial sensor data. We design methods that extract a motion's significant technical elements as well as methods that combine several measurable elements over time to extract motion features responsible for the aesthetic impression of a sports performance. In a neural network environment those feature extractors can then give the possibility to automatically evaluate and rank different performances in mobile training and competition systems, which could contribute to a better measurability and objectivity in performance-oriented sports as gymnastics and figure skating.

## 1 Introduction

With the increasing use of inertial sensors as motion capture tool, the number of available methods for processing and application of the sensor data has been increasing as well. Nowadays, it is possible to make valid assumptions on body parameters and a subject's behavior during a performance from the acceleration, angular rate and magnetic field sensor data, and to even determine body segment orientations and body joint positions with high accuracy. The gradual enhancement of inertial sensors and their processing methods is likely to lead to more professional sensor-based sports training and performance monitoring tools in the future: the cheap and light devices capture human motion in a direct way and do not depend on outer capture conditions so that they can also be used over a wide area as it often occurs in sport venues. Here, the main question is how to find specific motion knowledge from the sensor data, so that useful information can be extracted and provided to users like coaches and athletes. In the following, we particularly want to discuss how inertial sensors can be used to

© Springer International Publishing Switzerland 2016
P. Chung et al. (eds.), *Proceedings of the 10th International Symposium
on Computer Science in Sports (ISCSS)*, Advances in Intelligent Systems
and Computing 392, DOI 10.1007/978-3-319-24560-7_14

assess motion performances containing artistic elements that can generally only be evaluated by human vision.

One of the main issues in modern performance-oriented Olympic sports like gymnastics and figure ice skating is to assure an adequate level of objectivity in the judging and evaluation of a motion performance. Research shows that a judge's individual perception is generally biased, either unconsciously or on purpose, and that it is difficult to maintain objectivity when subjective human decision-making serves as the base of the ranking [9, 10]. Current scoring and judging systems of respective sports aim to assure a higher level of objectivity by reducing the influence of individual judging decisions on the final results. For this, they often include an additional measure as scalable or quantifiable performance evaluation factor. Inertial sensors can offer innovative motion assessment possibilities here and contribute to new credibility, but also understandability, for both athletes and spectators.

To make use of the raw multidimensional inertial sensor data, good strategies for the extraction of relevant motion information are necessary. Machine learning and database retrieval technologies offer a wast variety of algorithms for speech and multimedia processing. An overview of methods used in the context of sports data can be found in [1]. Many of them focus on low-level features and extract information directly from the raw sensor data, such as [6] and [3]. For the chosen evaluation task, we need to find reliable and valid features extractors that can also evaluate a performance with respect to aesthetic impressions. We therefore focus on higher-level features that contain semantic information like positional and temporal information of joints as well as relational information between body parts similar to the ones presented in [4]. We furthermore consider influences on a user's aesthetic perception as a combination of certain dominant semantic data relations that can be numerically determined over time. Similar problems exist for the parametrization of subjective music and video perception and are generally defined under the term computational media aesthetics [5]. We propose to then use a combination of those feature extractors in neural network similarity and proximity measures to obtain an output score that can be used as additional evaluation measure for the ranking in sporting competitions.

## 2   Data Processing

The common raw inertial sensor data is very sparse in the sense that it can only offer information on an object's angular velocities, the direction of gravity and the earth's magnetic field. Simple characteristics and anomalies of a motion performance can be found from the raw data with statistical measures or FFT and wavelets filters. Body segment orientations and joint positions, as they can be obtained with other motion capture technologies however, cannot be derived immediately and have to be computed in a post processing step to be used for a feature extraction. We estimate the orientations of body segments with a Complementary Filter based on discrete cosine matrices in quaternion representations [2]. It initially got developed for attitude reconstruction in robotics

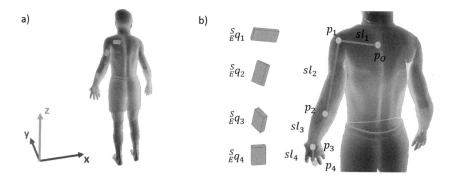

**Fig. 1.** a) Inertial capturing in the defined global coordinate system. b) FK-principle used in the data processing for the left arm.

and has been proven to be also very accurate, fast and applicable for sporting applications in previous experiments. Since the filter is based on quaternion representations, the resulting estimates are invariant to gimbal lock caused by singularities in the motion data. At every sample i, the orientation $_E^S q_{ji}$ for a body segment j in the sensor-earth coordinate frame is estimated by fusing the sensor data: a first estimate is obtained from the integration of the angular velocities in quaternion form and then refined by reference information on the error of the integration estimate from the accelerometer and magnetometer data. Using those orientation estimates, the end position $p_j$ of the segment j can then be computed in relation to the segment's origin $p_O$ under the principle of forward kinematics: with a known $sl_j$ representing the segment length of the performing athlete, $p_j$ is determined by the quaternion vector rotation $p_j =_E^S q_j \otimes \boldsymbol{\vartheta} \otimes_E^S q_j^*$, where the vector $\boldsymbol{\vartheta}$ is the three-dimensional segment length $[sl_j\ 0\ 0]$ and positional information depicted in the global coordinate system with z parallel to the direction of gravity and x referring to the lateral axis, see Figure 1.

## 3   Motion Feature Extractors

The strength and applicability of a mobile motion analysis tool depends largely on how well the processed inertial sensor data can describe a motion performance. Variables like the number and placement of the sensors used, the length and especially the type of the motion to be evaluated have an influence on the formation of general motion descriptions and should be taken into account for the creation of efficient and meaningful feature extractors, which is important for any database mining system: poor features can miss out on true hits, whereas too many or irrelevant features can lead to over selection and false hits. The idea of our work is to use universal assumptions on motion performances that can then be adapted for a use in a specific motion. A motion evaluation is subject of two general kinds of motion descriptors in our definition: features that extract technical determinants of a motion and features that determine the aesthetics of a motion to rate overall impression. The presented general feature extractors are illustrated by application examples for individual figure skating, which offers a balanced mix between technical and aesthetic motion parts and is regularly said to be a matter of conflict of interests in judging.

## 3.1 Technical Motion Features

$\mathcal{F}_T$ is the set of all feature functions that extract kinematic quantities or sequences of body movements within a performance under the principle of biomechanics. They can be computed for both performance- and result-oriented motions, as well as even target-oriented team sports. Therefore they can also be used for various tasks in mobile systems like motion analysis, motion classifications, and the intended evaluation.

1. **Body Segment Orientations** The estimated segment orientations $_E^S q$ give information on how specific body parts are oriented in the global space with respect to the Euler angles roll $\phi$, pitch $\theta$ and yaw $\psi$ around the motion axes x, y and z, whereas in many cases especially $\theta$ is of interest. One or more angles at certain instants of a motion build our first feature function $F_{T1}$. In figure ice skating for example, the inclination $\theta_{lT}$ at the elevated left leg measured from a sensor attached to the thigh is a relevant quality measure in charlotte spirals and camel positions, see Figure 2a). Another relevant orientation is the heading $\psi_{rF}$ at the (right) foot, which can help to detect cheated jumps with rotations not fully completed by the time of landing when the leg to land is the right one.

2. **Angles Between Body Segments** The difference between the orientations $_E^S q_{S1}$ and $_E^S q_{S2}$ of two neighboring body segments determine an angle $\delta_{\angle S1, \angle S2}$ at the connecting body joint like the bending of a knee or elbow. One or more of such joint angles at certain instants of a motion build this relational feature function $F_{T2}$. In figure skating, it can for example assess the sitting depth of a left-sided sit spin by the knee bend angle $\delta_{lT,lS}$ formed from the inclination angles $\theta_{lT}$ and $\theta_{lS}$ at thigh and shank, see Figure 2b).

3. **Joint Positions** The segment end positions $p_j$ give information on the x, y and z positions $x_j$, $y_j$ and $z_j$ of selected body joints in relation to the origin $p_O$ of their respective kinematic chain. One or more positions at certain instants of a motion build the feature function $F_{T3}$. In figure ice skating it can for example become possible to identify double footed landings with the position $z_{lF}$ of the right foot when the right leg is the leg to land as shown in Figure 2c); or to discover unwanted ice contact with the hands from the positions $z_{rF}$ and $z_{lF}$ of the finger tips.

4. **Relation Between Joint Positions** The distances $p_{j1,j2}$ between two joint positions (either within the same or a different kinematic chain) along x, y and z can be expressed in the spatial relations $x_{j1,j2}$, $y_{j1,j2}$ and $z_{j1,j2}$. One or more of such joint distances at certain instants of a motion build the relational feature function $F_{T4}$. Similar to $F_{T1}$, the positional difference $z_{rT,rS}$ between right thigh and shank can measure the spread in the hip

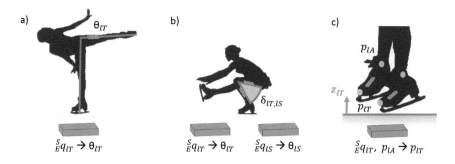

**Fig. 2.** Illustration of a possible use for the technical motion features a) $F_{T1}$, b) $F_{T2}$ and c) $F_{T3}$ and their proposed sample implementation.

joint in right-sided charlotte spirals and camel positions. $z_{rH,rS}$ and $z_{lH,lS}$ indicate a general low positioning of the wrists in relation to the shoulder joints.

5. **Temporal Course** All of the previous features cannot only be considered at certain instants of a motion, but also under their temporal aspects within the complete technical course of a motion. Often, it is more relevant to retrieve knowledge on the succession of rotational and positional aspects than on single instants, as for example the arm positions during a whole jump with takeoff, rotation and landing. One or more sequences of all previous technical measures build feature function $F_{T5}$. The temporal aspect of $F_{T5}$ then requires temporal processing methods in the following evaluation computation.

## 3.2 Aesthetic Motion Features

The set $\mathcal{F}_A$ is the collection of features that help to unveil aspects of a performance's impression and beauty, which is generally a matter of subjective perception. The idea in this work is to find several quantifiable factors (based on the information we obtain with the technical motion features) that impact the individual impression of an observer about a motion's aesthetic quality. Similar questions of aesthetic perception in music and video information retrieval already generated feature description strategies for non-motion multimedia data [8, 7], which we will use as a base for the design of our aesthetic motion features. In particular, they are based upon the parameters dynamics, flow, density, clarity and neighboring relations. Since aesthetic impression is generally formed with the progression of time over the course of a motion, we then come to propose the following feature designs.

1. **Motion Expression** The expression of a motion performance (generally also referred to as dynamics) can be described in several ways. One is by the range $rp_j$ of the position of certain body joints along $rx_j$, $ry_j$ and $rz_j$,

assuming that a larger motion range, and therefore a higher spatial coverage, is more impressive to the human eye. Another one is by the maximum angular velocities $mAV_s$ reached within a performance at a body segment, assuming that among two executions of the same motion, the one of higher angular velocity appears to be more dynamic. One or more temporal progressions of those measures build the feature function $F_{A1}$. The sequences $\{rz_{rH,1}, rz_{rH,2}, \ldots, rz_{rH,n}\}$ can for example rate the expression during a step combination of length n at the right wrist in up- and downward direction, $\{rx_{rH,1}, rx_{rH,2}, \ldots, rx_{rH,n}\}$ in left-right direction and $\{mAV_{rH,1}, mAV_{rH,2}, \ldots, mAV_{rH,n}\}$ on base of the angular velocities.

2. **Motion Flow** A motion performance is generally considered to be aesthetic if is it skillfully executed and follows a smooth flow, without sudden, unexpected events. Disturbances like sudden changes of directions and positions are very likely to be displayed in the data in form of irregularities, curbs or data peaks. We compute the positional difference of neighboring positions along all axes $sp_j$ and the rotational difference of neighboring quaternions at a certain joint $sr_j$ within a certain time span. Smooth motions should then not undergo any sudden large deviations from the mean $sp_j$ and $sr_j$ values of the selected time frame. Temporal progressions of one ore more of those difference measures form the feature function $F_{A2}$. Stumbling for example can result in sudden counter movements at the upper extremities that can then be retrievable as peaks in the sequence $\{sr_{H,1}, sr_{H,2}, \ldots, sr_{H,n}\}$ or $\{sl_{H,1}, sl_{H,2}, \ldots, sl_{H,n}\}$ of length n.

3. **Performance Density** Counting the number of events $nE_j$ that occur within a certain time frame at a specific body part gives us a density measure. One or more event counts then build the feature function $F_{A3}$, which is difficult to define as it is depending largely on the specific requirements of a sport and the specification of an event. First, it is necessary to determine events that give a positive impression of skill and strength to an observer. Then, it is necessary to classify such events from the motion data to be able to detect their number of occurrences $\sum nE_P$. Events that could be detected easily from the sensor data in figure skating are longitudinal turns in spins within a defined time frame n by the changing of heading $\psi$ of a sensor attached to the athlete's pelvis as $E_P = \sum_{t=1}^{n} \psi_t$ or the number of jumps within a defined time frame n by the number of times the acceleration at the pelvis surpasses a certain threshold $AccT$, indicating the landing motion of a jump, as $E_P = \sum_{t=1}^{n} AccT_t$.

4. **Motion Clarity** The clarity of a motion performance shall be determined from the number of side motions that occur during a movement at a certain joint or body segment that is not directly related to the main performance, as for example the motion of the arms during a spin. This feature function $F_{A4}$ is built by the sum of one or more translations in the x,y and z axes or rotational changes at body parts that are not directly involved in a specific movement. The easiest way to describe such sequences is by motion jitter $\sum_{t=1}^{n} exAV_{rH,n}$ as the accumulation of extreme points $exAV$ (and hence data peaks) in the respective angular velocities during the time frame n.

# 4 Evaluation Computation

We have introduced algorithms to extract features that represent both a motion performance under technical and aesthetic aspects from the inertial sensor data and combine them into a feture set $\mathcal{F}$ for the respective motions. In the following, it is now necessary to define measures that can score or rank different performances on the base of this retrieved information. In general, a performance will always be subject to individual motion styles that are amongst others influenced by physical differences in athletes' body properties, so that it is not possible to make assumptions on the athlete-independent ideal motion. We therefore propose to evaluate motion performances amongst each other, with respect to the feature descriptions extracted at certain key phases for $\mathcal{F}_T$, and over the whole course of a certain performance for $\mathcal{F}_A$, and to then generate a top-to-bottom ranking on this base, with an additional score about how close two motions performances have been to each other. For this, we need to define measures that determine similarities between the feature descriptions at such key phases and their temporal motion evolutions, and that numerically grade deviations between each other in an appropriate way. Proximity between features and media data can be computed in several different ways, using methods of neural networks, such as classification trees, k-Nearest Neighbour (k-NN), Bayesian classifier, rule-based approaches or Support Vector Machines (SVMs). Suitable measures should be investigated and discussed in more detail in consecutive work.

# 5 Results

For the completion of an evaluation system, the designed feature extractors need to be tested and appropriate rating values be determined that can score deviations and similarities in a neural network system. Along with this, it is essential to find clear rules and annotations of certain movements to be able to apply the proposed aesthetic motion feature extractors, as the main difficulty here is that often aesthetic knowledge can be formulated for specific cases, but not be easily learned for an automatic and machine-based retrieval.

Experiments will be conducted in summer and within the rest of the year to build a test and training database for the retrieval system, and to compare the results of the designed evaluation and scoring measure to the rankings of human judges. First results with practical data are intended to be presented at the conference.

# 6 Conclusion and Outlook

In this paper we have discussed how main features of a motion performance can be extracted with respect to their semantic aspects for both motion technique and aesthetic impression. We discussed technical motion features based on biomechanical performance determinants as well as innovative aesthetic motion features that are derived from related multimedia emotion retrieval strategies.

Especially the proposed aesthetic motion features are of high interest, since they are very relevant for future applications in case of a successful and appropriate technical implementation and could therefore lead the way towards a new level of objectivity in sports that are prone to judging bias in future. However, they are also very difficult to develop in such a way that they can be used automatically and independently in mobile applications. Here, we demonstrated the use of different feature extractors under the example of individual figure ice skating, but all features can be used with different sports in a similar way.

In the next steps, it is now necessary to collect various experimental motion data to assess the functionality of the proposed motion features under different conditions, sports and motion types and to develop the evaluation measures that can give an output scoring result. Then, we will be able to evaluate the use of the complete developed system and enforce probable steps for the application in mobile systems.

# References

1. Baca, A.: Methods for recognition and classification of human motion patterns–a prerequisite for intelligent devices assisting in sports activities. MATHMOD 2012, February 15-17, Vienna, Austria (2012)
2. Euston, M., Coote, P., Mahony, R., Kim, J., Hamel, T.: A complementary filter for attitude estimation of a fixed-wing uav. In: Intelligent Robots and Systems, 2008. IROS 2008. IEEE/RSJ International Conference on. pp. 340–345 (Sept 2008)
3. Harding, J., Mackintosh, C., Hahn, A., James, D.: Classification of aerial acrobatics in elite half-pipe snowboarding using body mounted inertial sensors (p237). In: The Engineering of Sport 7, pp. 447–456. Springer Paris (2008)
4. Helten, T., Brock, H., Müller, M., Seidel, H.P.: Classification of trampoline jumps using inertial sensors. Sports Engineering 14(2-4), 155–164 (2011)
5. Nack, F., Dorai, C., Venkatesh, S.: Computational media aesthetics: finding meaning beautiful. MultiMedia, IEEE 8(4), 10–12 (Oct 2001)
6. Pansiot, J., Lo, B., Yang, G.Z.: Swimming stroke kinematic analysis with bsn. In: Body Sensor Networks (BSN), 2010 International Conference on. pp. 153–158 (June 2010)
7. Yang, Y.H., Lin, Y.C., Su, Y.F., Chen, H.: A regression approach to music emotion recognition. Audio, Speech, and Language Processing, IEEE Transactions on 16(2), 448–457 (Feb 2008)
8. Zhang, Y., Zhang, L., Zimmermann, R.: Aesthetics-guided summarization from multiple user generated videos. ACM Trans. Multimedia Comput. Commun. Appl. 11(2), 24:1–24:23 (Jan 2015), http://doi.acm.org/10.1145/2659520
9. Zitzewitz, E.: Nationalism in winter sports judging and its lessons for organizational decision making. Journal of Economics and Management Strategy 15(1), 67–99 (2006), http://dx.doi.org/10.2139/ssrn.319801
10. Zitzewitz, E.: Does transparency reduce favoritism and corruption? evidence from the reform of figure skating judging. Journal of Sports Economics 15(1), 3–30 (2014)

# Part IV
# Modelling and Analysis

# The Aikido inspiration to safety and efficiency: an investigation on forward roll impact forces

Andrea Soltoggio[1], Bettina Bläsing[2], Alessandro Moscatelli[3], and Thomas Schack[2]

[1] Computer Science Department, Loughborough University, Loughborough, UK
[2] Neurocognition and Action Research Group, Bielefeld University, Bielefeld, Germany
[3] Cognitive Neuroscience, Bielefeld University, Bielefeld, Germany

**Abstract.** Aikido is a Japanese martial art inspired by harmony and intelligent exploitation of human body movements, a consequence of which is believed to be a minimisation of impacts. This study measures the effectiveness of aikido-specific movements to minimise impact forces, and arguably the risk of injuries, in person-to-floor contact. In one experiment, we measured a significant reduction of impact forces with the ground for aikido experts during a forward roll in comparison to untrained participants. This first initial result encourages further studies of aikido techniques in areas such as safety and efficacy in sport exercise, safety during full body motion involving falls and impacts, transfer to human-robot interaction and training of elderly people.

## 1 Introduction

The aikido philosophy and practice teach harmony and the understanding of human movements to induce unity and agreement between the practitioner and his/her surroundings, including a partner [13, 10, 11]. The techniques studied in aikido allow for an intelligent reaction to the different attacks performed by the partner. Aikido techniques aim not only to neutralise the attack, but also to restore equilibrium and harmony between the athlete attacking (*uke*) and the one reacting (*tori*). As described in [6], aikido "transforms aggression into cooperation": an idea that has seen aikido being the subject of a number of sociological studies [7, 4] looking at mediation and conflict resolution.

Aikido concepts like *harmony*, *unity*, and *agreement*, in the sense of a continuous dialog with the movement of the partner to manage his/her action force, suggest a collaborative and intelligent attitude in the resolution of potentially dangerous dynamic movements. Frequent approaches to resolve impact conditions are based on strength, a way to further disrupt harmony and unity, and thus are to be avoided in aikido. An intelligent resolution of dynamic conflictual conditions, on the other hand, involves early understanding of the nature of the movement, and the application of techniques that mitigate the risk or intensity of the impact. This "intelligent attitude" is often refined to the highest technical and philosophical standards by the highest ranking aikido masters. The

P. Chung et al. (eds.), *Proceedings of the 10th International Symposium on Computer Science in Sports (ISCSS)*, Advances in Intelligent Systems and Computing 392, DOI 10.1007/978-3-319-24560-7_15

120                                                          A. Soltoggio et al.

effectiveness of a large spectrum of ideas, philosophy and practice, when they converge to result in the minimisation of impact forces, could be evaluated in relatively narrow but measurable way, i.e. by means of precise measurements of impact forces, which is what we aim in this study.

In aikido, when a person experiences a fall, the ground is seen as a partner, not to be avoided or aggressively confronted, but rather to be met with what can be defined as "friendly" reactions and movements. The initial conditions that determine a fall have to be detected as early as possible to prepare for an appropriate response. Such a response involves the execution of precise and technical movements by the aikido practitioners, resulting in a well executed *ukemi*. In most cases, a fall can be converted into a low-impact sequence of movements resulting in a forward or backwards roll.

Most aikido schools put considerable emphasis and technical detail in the execution of *ukemi*. In this study, we refer to the school of Katsuaki Asai Sensei, a Japanise aikidoka who trained directly with the founder of aikido, Morihei Ueshiba between 1955 and 1965 [2]. Figure 1 is a sequence of photos taken from a video of Asai Sensei performing a forward roll. It can be noticed that the contour

**Fig. 1.** Katsuaki Asai Sensei performing a forward roll [1]. In the snapshot 2 and 3, a dashed circle is added to indicate the circular shape of the body that facilitates smooth low-impact ground contact.

of the body assumes a circular form during the execution of the roll (sequence 2 and 3), as outlined by the overlying dashed circle. The full sequence in the video conveys a natural, smooth and low-impact sequence of movements. Asai Sensei, who founded Aikikai Deutschland when he moved to Germany in 1965, taught instructor Thomas Gertz, third Dan, who provided technical support and demonstrations to conduct this study.

## 2 Method

A force sensor integrated with the floor, measuring 90x90cm, was used to measure horizontal and vertical forces that participants exerted on the ground when performing a forward roll. Additionally, video cameras and a Vicon system of 12 infrared cameras (for tracking position of body parts in 3D space) were employed to capture additional data for a subset of participants. Figure 2 illustrates the setup.

**Fig. 2.** Experimental setup. A force platform is incorporated at floor level (here hidden under a protective foam mat of 7mm). Thicker safety mats (2cm) are placed around the platform to protect participants. The Vicon system was used to track positions of some key body elements, but due to occlusions and unreliability of the tracking was not used for the analysis. Instructor Thomas Gertz is showing the initial position of the movement.

The forward roll is a dynamic body-to-floor contact movement that may approximate a gentle type of fall. A realistic type of fall would involve an initial standing position, although not necessary completely upright, as in the example of Figure 1. Unfortunately, this movement, although it appears natural and simple when performed by an expert, cannot be safely tested on untrained participants without risking injuries. For both pedagogic and safety reasons, instructor Gertz teaches a range of increasingly challenging rolling movements, involving starting positions from completely crouched on the ground, to kneeling and half standing. In this way, beginners can start practicing rolls from a low and safe position. Experts may also perform rolls from low positions to refine their technique with slow and precise movements. Preliminary tests revealed that rolling from a kneeling position (starting position as in Figure 2), and using a 7mm foam mat to cover the force sensor on the ground, was a sufficiently safe movement if performed under the supervision of an expert.

The precise movement, performed by instructor Gertz, was video recorded and demonstrated to participants who were asked to perform the same movement. Figure 3 shows a sequence of photos during the demonstrated forward roll. Interestingly, rolling froward from a kneeling position can be hardly seen as an

**Fig. 3.** Instructor Thomas Gertz, third Dan Aikikai Deutschland, demonstrates a forward roll from a low kneeling position. The movement starts from a static position with both feet and knees on the ground. The movement can be performed in two symmetrically equivalent variations: lowering at first either the left or the right arm/shoulder. Impact forces are registered and measured during phases 4 to 7.

impact-inducing movement. However, as the results will reveal, even such a low height movement can be challenging for untrained participants.

**Participants.** The experiment involved N=7 trained participants with 6 months to 10+ years of regular aikido training. Additionally, untrained participants (N=11, 6 males, 5 females) with no previous experience in aikido were tested. Nine out of the 11 untrained subjects performed regularly other sports, and all were selected among healthy subjects capable of performing basic physical exercise.

**Procedure.** Each participant was demonstrated the movement in Figure 3 and invited to observe it in order to reproduce it. In particular, the participants were instructed to following these steps: 1) assume the correct initial position; 2) lower one shoulder (either left or right) towards the ground to prepare the roll; 3) engage in the dynamic movement trying to reach a static position as in phase 8 of

Figure 3. After the explanation, the participants had the opportunity to practice a few times, under the supervision of one expert, until he or she felt comfortable and capable of executing the sequence of movements correctly. Because the goal of the movement was that of performing a roll within the measuring area (Figure 2) and of ending the roll as in Figure 3 (phase 8), participants practiced until they could reach that objective, taking between 2 to 6 preparatory rolls.

**Measurements.** After the practice rolls, each participant had four rolls recorded, two starting by lowering the left shoulder, and two starting with lowering the right shoulder. Finally, the static vertical force (i.e. equivalent to the weight) was measured for each participant in order to compute the ratio of dynamic over static vertical forces.

# 3 Results

Figure 4 shows the impact forces recorded by the force platform during rolls performed by two participants with 3 and 10 years of aikido experience (panels A and C respectively), and two participants with no previous training (panels B and D). There appears to be little difference between trained subjects (panels A and C) despite difference in history and years of training. Untrained subjects instead appear to have high peaks of vertical forces. We applied a Linear Mixed Model (LMM) [3] to test for the difference in the force ratio between trained and untrained participants (Figure 5). Mixed models distinguishes between *fixed-effect* predictors accounting for the effect of the experimental variable (i.e., years or practice) and *random-effect* predictors accounting for idiosyncratic differences between participants. We applied the following model:

$$\boldsymbol{y}|\boldsymbol{b} = b \sim \boldsymbol{X}\beta + \boldsymbol{Z}\boldsymbol{b} + \boldsymbol{\epsilon}, \tag{1}$$

Where $\boldsymbol{y}|\boldsymbol{b} = b$ is the force ratio in participant $b$, $\boldsymbol{X}$ is the years of aikido practice, $\boldsymbol{Z}\boldsymbol{b}$ is the random-effect predictor and $\epsilon$ is the error term. The estimated coefficient $\beta$ was $-0.1 \pm 0.01$ ($\beta\pm$ Standard Error), which means that the force ratio decreases on average of 0.1 unit per year of practice. The effect of year of practice was statistically significant (Likelihood Ratio Test, $p < 0.001$).

We also tested an alternative LMM where the level of expertise was coded as a categorical predictor. Participants were divided into two groups, non-expert (less than two years of practice) and expert (two years of practice or more). The effect of group was statistically significant ($p < 0.001$) confirming that practice of aikido significantly reduce the contact force during rolling.

In two cases, untrained subjects recorded forces higher than 3.5 times their body weights, a surprising factor considering that the movement starting position (Figure 3) is at a low height, with both knees on the ground. Given such high impact forces registered by untrained subjects, it was decided that a roll starting from a higher position of the body centre of mass could not be safely performed by untrained subjects. Nevertheless, it was deemed interesting to compare the

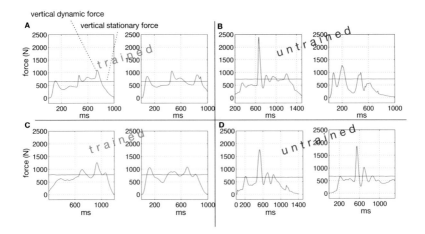

**Fig. 4.** Impact forces measured during rolls performed by two trained participants (one expert, one intermediate), and two untrained participants. The vertical force is reported on the vertical axis, time in ms is on the horizontal axis. Each panel (ABCD) shows two rolls for one participant initiating the roll with the left arm (left graph) and with the right arm (right graph). (A) Expert participant's rolls (3 years of aikido practice). (B) Untrained participant's roll. (C) Expert participant's roll (10 years of aikido practice). (D) Untrained participant's roll.

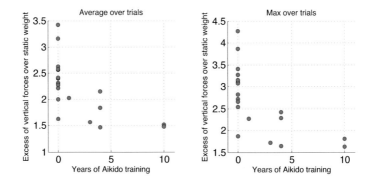

**Fig. 5.** Ratio of dynamic forces over stationary forces for all subjects taking part in the experiment plotted against the aikido experience (in number of years).

previous data with a sample of measurements involving a roll movement from a standing position, i.e. reproducing the movement in Figure 1, performed exclusively by trained subjects. Figure 6 shows two rolls from a standing position

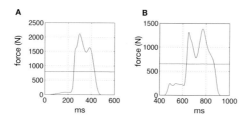

**Fig. 6.** Vertical forces during two rolls (performed by two different subjects) from a standing starting position. Compared to rolls from a low height (Figure 4, which had a duration of approximately 800ms, these movements from standing have a duration of 200 to 300ms, i.e. there are executed at approximately 3 times the speed of the low-height rolls. Trained subjects are nevertheless capable of minimising impact forces in this case as well, resulting in impact forces to weight ratio that are still lower than those recorded by untrained subjects starting from a ground position.

performed by experts with 3 and 4 years of aikido practice. The peaks of forces are in these cases lower than those registered by untrained subjects when rolling from a kneeling position. In other words, for a trained subject, falling from a standing position, and transforming the fall into a roll, measures less impact forces than for the average person rolling over starting already from the ground.

## 4    Discussion

In this study, we measured ground impact forces during forward rolls performed by aikido experts and amateurs as well as novices. We found that peak impact forces in the group of novices was higher than in the group of aikido-experienced participants. Furthermore, the variance of peak vertical forces was smaller in the group of aikido practitioners, suggesting that the movement was performed more homogeneously over trials by this group than by the novice participants.

It has to be stated that this pilot study with seven aikido practitioners of different expertise level (two experts with 10 years of training, five amateurs practicing on intermediate level, 1 to 4 years) and eleven novices does not have the potential to provide a full picture regarding force compliance in aikido rolling techniques. However, as this is the first study of this kind, we are presenting substantial evidence that aikido techniques seem to reduce impact forces during a simple roll movement. The results of this preliminary study need further substantiation on the basis of measuring impact forces from participants of different expertise level performing different rolling techniques (e.g., from different starting positions). Looking at individual participants' data, large differences, particular among untrained participants, were observed. A venue of future research should be to improve our understanding on why some participants recorded very high impact forces (thereby being more likely to be injured) while others recorded rather low impact forces.

In this study, the use of the Vicon system was intended to further analyse the precise dynamics of each roll, which is instrumental to gather more insights on the precise techniques and differences among participants. As markers were often occluded or lost during the rolls, future studies should employ flatter or softer markers, securely attached, and with a high number of cameras. This will enable a technical or pedagogical description of the movement, e.g. as 3D reconstruction of the rolls that minimise impact forces.

Results of our study clearly point at the potential of aikido training to reduce ground impact forces during rolling, a topic that certainly deserves further investigation. Our findings promise to be relevant not only for martial arts training, they also bear high relevance for sports training in general and for all contexts in which the reduction of ground impact forces is desirable and important for reasons of security, to reduce injury and damage. Learning to fall or roll smoothly with low impact can be a protective measure in many sports (e.g., [8]) as well as in everyday situations, particularly for the elderly. [12] showed that even after a short martial arts fall training of 30 minutes, young adults without experience of martial arts training were able to reduce hip impact forces and velocities in martial arts falls from a reduced height (kneeling), and [5] achieved comparable results with elder adults (60-80 years) after five training sessions. Finally, similar strategies are relevant to bipedal humanoid robotics, particularly in robot soccer or other real-world applications, as minimizing impact forces from falls is crucial to maintain the robots functional in situations in which falls cannot be avoided [9].

## 4.1   Conclusion

This study analyses one particular full body rolling movement performed by both aikido experts and untrained participants. Employing a force platform that measures impact forces with the ground, it was possible to measure a significant difference of impact forces between aikido experts and untrained participants. The study contributes to validate the hypothesis that aikido movements seek low impact dynamics. Not only are impact forces reduced when aikido techniques are applied, but arguably, the movement is performed in a more efficient way, reducing the energy dissipated on the ground by impact forces. The results also suggest that aikido techniques to control falls could be beneficial to reduce the risk of injury for particular participants, e.g. in particular sports, or for elderly people that are more prone to falls.

# References

[1] Asai, K.: Youtube: Katsuaki Asai Allemagne [last checked june 2015] (1970), https://youtu.be/r8OCfEn77Ts

[2] Asai, K.: Interview with Katsuaki Asai [last checked june 2015] (1993), http://www.aikidojournal.com/article?articleID=309

[3] Bates, D., Mächler, M., Bolker, B., Walker, S.: Fitting linear mixed-effects models using lme4. arXiv preprint arXiv:1406.5823 pp. 1–51 (2014)

[4] Edelman, A.J.: The implementation of a video-enhanced aikido-based school violence prevention training program to reduce disruptive and assaultive behaviors among severely emotionally disturbed adolescents. (1994)

[5] Groen, B.E., Smulders, E., De Kam, D., Duysens, J., Weerdesteyn, V.: Martial arts fall training to prevent hip fractures in the elderly. Osteoporosis international 21(2), 215–221 (2010)

[6] Kroll, B.: Arguing with adversaries: Aikido, rhetoric, and the art of peace. College Composition and Communication pp. 451–472 (2008)

[7] Saposnek, D.T.: Aikido: A systems model for maneuvering in mediation. Mediation Quarterly 1987(14-15), 119–136 (1986)

[8] Shuman, K.M., Meyers, M.C.: Skateboarding injuries: An updated review. The Physician and Sportsmedicine (0), 1–7 (2015)

[9] Ruiz-del Solar, J., Moya, J., Parra-Tsunekawa, I.: Fall detection and management in biped humanoid robots. In: Robotics and Automation (ICRA), 2010 IEEE International Conference on. pp. 3323–3328. IEEE (2010)

[10] Ueshiba, K., Ueshiba, M., Stevens, J.: Best aikido: the fundamentals. Kodansha (2002)

[11] Ueshiba, M.: The art of peace. Shambhala Publications (2002)

[12] Weerdesteyn, V., Groen, B., van Swigchem, R., Duysens, J.: Martial arts fall techniques reduce hip impact forces in naive subjects after a brief period of training. Journal of Electromyography and Kinesiology 18(2), 235–242 (2008)

[13] Westbrook, A., Ratti, O.: Aikido and the dynamic sphere: An illustrated introduction. Tuttle Publishing (2001)

# To evaluate the relative influence of coefficient of friction on the motion of a golf ball (speed and roll) during a golf putt

Dr Iwan Griffiths[1], Rory Mckenzie[1], Hywel Stredwick[1] and Dr Paul Hurrion[2]

[1]College of Engineering
Swansea University
Swansea, United Kingdom

[2]Quintic Consultancy
Sutton Coldfield, United Kingdom

**Abstract.** Pace control and green reading have been highlighted as important aspects of a golfer's arsenal when it comes to putting. The purpose of this study was to compare the ball roll characteristics across eight different surfaces and more specifically analyse the distance it takes for a golf ball to achieve pure rolling motion. Two different methods of collecting putting data were used during this study; a putting robot and a human subject. For each surface thirty putts were tracked for the first 40cm of their travel, namely; a putting mat, rubber, MFC, compact carpet, brass, PTFE, MDF and nitrile rubber, via a high speed camera (360 frames per second). A numerical model was used to determine the mean coefficient of kinetic friction from each of the eight surfaces. Results showed that the surface with the highest coefficient of friction ($\mu = 0.40$), allowed the ball to enter true roll at the earliest stage (3.94"), whereas the surface with the lowest coefficient of friction ($\mu = 0.11$), allowed the ball to enter pure roll at the latest stage (16.77"). Knowledge of this negative relationship may have great significance for golf coaches, players who are looking to improve their green reading skills, and golf course designers/greens keepers who are trying to further understand and improve putting greens.

**Keywords:** Golf putting; Roll; Skid; Coefficient of Friction

## 1   Introduction

Between 43% [1] and 45% [2] of all golf shots are played using a putter, which shows its importance within the game of golf. Pelz [1] measured inconsistencies of golf greens and found that there was a difference of 43% (73% vs 30%) in the number of balls that rolled into the hole when released at a constant velocity and direction prior to and after tournament play, while another study conducted in 2008 [3], dismissed ball roll as a controllable factor due to green inconsistencies'. This indicates that green inconsistencies often as a result of spike marks, are a large contributing factor in successful putts. However, Swash [2] stated that achieving rolling motion immediately upon striking the golf ball significantly reduces the negative effect of poor green surface conditions. Furthermore more recent research [4] suggests that a putt that rolls

© Springer International Publishing Switzerland 2016
P. Chung et al. (eds.), *Proceedings of the 10th International Symposium on Computer Science in Sports (ISCSS)*, Advances in Intelligent Systems and Computing 392, DOI 10.1007/978-3-319-24560-7_16

for a greater ratio of its length is more desirable for distance control. This may be due to the fact that the coefficient of rolling friction is significantly lower than that of sliding friction [5,6]. Therefore it should be in a golfer's interest to aim to achieve pure rolling motion as soon as possible during a putt.

Numerous studies [7,8,9] outline the principles of static and kinetic friction and how these have an effect on golf putting, however no research has looked at the effect the surface has on the rolling characteristics of a golf ball. Green reading has consistently been found to be the biggest contributor to overall putting performance [1,3,10]. These studies outline the importance of green reading, and suggest that more practice time should be given to this area of the putting game. However, although these authors suggest that golfers should look to improve this aspect of their game, there are very few practical implications that golfers can use to enhance this area.

It has been suggested [11] that stimpmeter readings can be used to measure the amount of friction that is present on the golf green. However, the stimpmeter only assesses the amount of rolling friction that is present on the green and does not take into account the initial skid phase that occurs at the start of the putt. Therefore a method that is more specialised to a golf putt is needed to determine the amount of sliding (kinetic) friction that is present on different surfaces, which may enable golfers and coaches to better understand why a golf ball may react differently on different types of surface. Therefore this study aims to determine the coefficient of kinetic friction of various types of surface, before analysing the effect that they each have on the ball roll characteristics of a putted golf ball (i.e. distance/time taken for the ball to reach true roll).

## 2      Method

This study used two ways of collecting putting data; a 'robot' method and a 'human' method. In the 'robot' method, a putting robot (based at Quintic Consultancy's golf laboratory) struck the golf ball in a tightly-controlled putt. The 'human' method used a real golfer to undertake a similar putt, but the experimental conditions were necessarily more variable. Data collection for the human method was conducted at Swansea University. The Quintic putting robot was set up to simulate a putt with an initial ball speed of 3mph, with a straight swing path to ensure a square club face during impact (See figure 1). For the human method, the golfer was also instructed to aim to achieve a ball speed of 3mph. An Odyssey (Callaway Golf Europe Ltd., Surrey, UK) White Hot #3 putter (putter length - 34"), with a 69° lie and 2.5° loft, was used for both methods (with the lie of the putter referring to how flat the putter is to the ground, and the loft referring to the angle of the club face). Five different Titleist Pro V1 golf balls were used during the study. These golf balls had a spherically-tiled 352 tetrahedral dimple design. All golf balls had three small black dots embedded on the side of the ball to allow the camera to identify and track the ball during experimental trials (see figure 2). These dots were formed with use of ink and did not affect the mass of the golf ball. The dots were aligned in a right angled triangle to calculate roll/spin. The Quintic ball roll (360fps USB 3) camera (see figure 3) was used to record each putt.

The camera tracked the putter for 4 inches prior to impact of the ball and the ball for the first 16 inches of the putt.

Thirty putts were collected for eight different surfaces namely; a putting mat (the classic original welling putt mat), 3mm thick rubber, 3mm thick melamine faced chipboard (MFC) and 5mm thick compact carpet, all used for the robot method. Naval brass (59% copper, 40 % Zink, and 1% tin), PTFE, 3mm MDF and nitrate rubber were used for the human method.

In order for a putt to be selected for analysis it had to meet the 'Quintic recommendations' for a 'valid putt'. The 'Quintic recommendations' suggest that ideally the initial launch angle of the ball immediately after it leaves the club face should be between 0.75° and 2.5°. The cut and hook sidespin on the ball was not allowed to exceed 20 RPM [12]. From these recommendations putting parameters were set to allow 'valid putts' to be collected for analysis (table 1).

**Table 1.** Putting parameters for the putting robot and human subject.

| Putting parameters | Quintic putting robot | Human subject |
|---|---|---|
| Impact ball speed | 3mph ± .25 | 3mph ± .5 |
| Initial launch angle | 2° ± .5 | 0° - 3° |
| RPM cut or hook spin | 0 - 20 | 0 - 20 |

**Fig. 1.** Set-up of the Quintic putting device

Even with the use of the Quintic putting robot, it was very difficult to produce exactly the same putting parameters (i.e. initial launch angle, impact ball speed and RPM cut or hook spin) for each different putt. This was due to the human error in the pull back and release phase, where the putting device was pulled back to the distance required (40cm) for the correct speed (3mph) before being released. It was even more difficult to reproduce the same putting parameters for the human method due to the variability in players putting stroke. Therefore achievable putting parameters were set to assure that all of the putts collected for analysis were as similar as possible, while meeting the 'Quintic recommendations' for a 'valid putt'. The ranges were set so that they were as small as they possibly could be, to assure that the experiment was as accurate as possible. The ranges for the impact ball speed and initial launch angle for the human subject had to be increased further (see table 1) due to the difficulty in reproducing the same putt each time, which resulted in reduced accuracy for this method.

## 2.1 Additional information/definitions

- Impact ball speed: is the speed of the golf ball immediately after it leaves the clubface.
- Initial launch angle: is the launch angle of the ball immediately after it leaves the putter face.
- RPM cut or hook spin: Is the amount of cut or hook spin acting on the ball after it has been putt (i.e. sidespin).

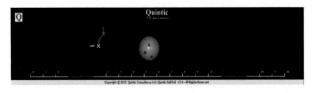

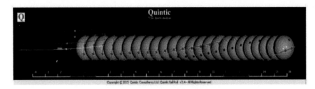

**Fig. 2.** A visual representation of how the golf ball is tracked during a putt

## 2.2 Set-up of Quintic robot device

The Quintic putting robot (see figure 1) was set up to take a right handed putt with the following putting parameters; Path: square to square, shaft: vertical, arm hang: 20°, lie: 70°. The robot was set up 90 degrees to the left of where all putts were to be taken. The putter was placed in the clamp of the robot and a spirit level was used to determine when the shaft was vertical before the clamp was tightened. These putting parameters were used to ensure that the putter had a square club face at impact.

## 2.3  Set-up of camera and alignment with the target

The maximum frame speed for the USB camera was 365.51 fps (frames per second); therefore the closest value to this (360 fps) was selected. The camera lens used was a fixed lens (12mm) with manual focus and exposure. The lens was adjusted until the three spots on the ball could be seen clearly, with sufficient contrast.
A mark was placed on the putting surface to denote the starting position of the ball, and another mark 8ft away denoting the hole/target. A 'T Bar' was then positioned

with the centre hole of the alignment tool positioned on the mark on which the ball was sat. A laser was positioned so that the beam ran along the centre of the 'T Bar' and through the mark denoting the target. A high speed camera (360 fps) was positioned 112cm (from camera lens) adjacent to the starting position of the golf ball. The camera exposure was calibrated within the 'Quinitc Ball Roll' software.

### 2.4    Ball and putter calibration

Two balls were positioned in the cut-outs in the 'T bar', ensuring that the 3 marked dots on each ball were in full view of the camera. Minor adjustment of the camera alignment was achieved by adjusting the feet of the camera, thus ensuring the two balls aligned themselves accurately with the yellow circles displayed on the computer screen and that the horizontal line aligned itself with the base of the 'T Bar'. The camera was then focused to ensure a clear image for analysis. The 'T Bar' was then removed to allow the calibration process to commence.

Four marking dots were applied to the putter, two at the bottom of the shaft and 2 at the toe of the putter for camera tracking purposes. The 'T Bar' was re-positioned on the ball mark (rotated 90 degrees). A laser was positioned such that it ran through the centre of the 'T Bar' and through the centre of the hole. The putter face was placed flush to the back of the 'T Bar' and was held so it was square to the target line. With use of a putter spirit level the club shaft was positioned such that the spirit level read the club shaft as vertical. When the four markers were identified on the putter, calibration of the putter was complete. A measurement of the depth (toe to middle of shaft) of the putter was taken and was input into the software when prompted.

### 2.5    Performing a putt

A ball was positioned on the marked starting position, with the arrow on top of the ball pointing towards the hole, and the three dots facing towards the camera. When the three dots on the ball had been identified by the camera a putt could be performed. For the Quintic putting robot, the putting device was drawn back (40cm) before being released. This would allow for the initial ball speed of the ball to be approximately 3mph after impact. For the human subject a normal putt was performed, with the subject aiming to achieve an initial ball speed of 3mph. After each putt had been performed, the data was viewed (within Quintic Ball Roll) to see whether a valid putt had been produced. If the putt was deemed valid it was saved for analysis. This process was done until 30 valid putts for each putting surface had been collected.

### 2.6    Recording of the putt

After each putt, putter and ball data were automatically shown both numerically and graphically via the computer screen providing data on: Impact club speed (mph), Pre-impact club speed, Attack angle (°), Impact ball speed (mph), Launch angle (°), Cut & Hook Spin (rpm), Initial ball roll (rpm), Start of forward rotation (inches), Distance to true roll (inches) and Time to true roll (seconds).

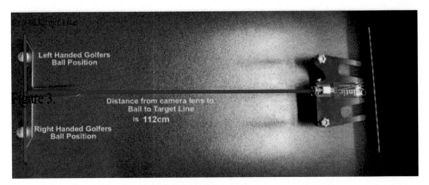

**Fig. 3.** Setup of the 'Quintic Ball Roll' system [12]

## 2.7    Determining the coefficient of friction

The mean coefficient of friction of each surface was determined with the use of a specialized excel spreadsheet [13], that was originally used to determine the sliding coefficient of friction of a bowling ball. This was adapted for use in golf putting. The golf ball was modelled as a sphere of homogeneous density of the appropriate size (1.15 g/ml) and mass (0.046kg) for a golf ball. The static and kinetic coefficients of friction were set to fixed values before simulation started. The ball usually starts off sliding, and then gradually begins to roll. This emerges from the simulation. The friction parameter is adjusted until the simulation matches the experimental result. A customised graph was used to estimate the static and kinetic friction coefficients of friction for each putt. The simulation starts with the angular speed of the ball set to zero and a given linear velocity. The linear velocity and estimated static and kinetic coefficients of friction of each individual putt were input into the system separately. The simulation process used numerical integration of the equations of motion, solving for both the linear and angular speeds of the ball. The distance required to achieve true roll was obtained from the simulation. This value was compared to the actual value that true roll occurred for the putt that was being analysed, which was obtained from the experiment. If these values matched (to 2 decimal places), the coefficient of friction was recorded. This procedure was done until a coefficient of friction value had been obtained for each putt (30 putts) of each surface. When this had been achieved the means and standard deviations of each surface were calculated within a separate excel file.

## 3 Results

The highest coefficient of kinetic friction for the robot method was obtained using the rubber surface ($\overline{\mu}$ = .29, $\sigma$ = .01), whilst the lowest was MFC ($\overline{\mu}$ = 0.11, $\sigma$ = 0.01). The distance to zero skid reflected this with the respective surfaces generating distances of 6.00 and 16.77 inches.

The highest coefficient of kinetic friction for the human method was obtained using the rubber surface ($\overline{\mu}$ = 0.40, $\sigma$ =0 .04), whilst the lowest was wood ($\overline{\mu}$ = 0.20, $\sigma$ = 0.02). The distances to zero skid were 3.94 and 9.45 inches respectively.

**Table 2.** Ball roll characteristics of eight different surfaces. Data are mean ± SD.

| Surface | Coefficient of Friction (μ) | | Mean distance to true roll. Inches (metres) |
|---|---|---|---|
| | Mean ± SD | Range | |
| **Robot** | | | |
| Putt mat | .24 ± .01 | .21 - .26 | 7.03 (.18) |
| Rubber | .29 ± .02 | .27 - .34 | 6.00 (.15) |
| MFC | .11 ± .01 | .09 - .13 | 16.77 (.43) |
| Carpet | .17 ± .01 | .15 - .19 | 10.33 (.26) |
| **Human** | | | |
| Brass | .29 ± .04 | .24 – .39 | 5.97 (.15) |
| PTFE | .34 ± .06 | .22 – .49 | 5.12 (.13) |
| Rubber | .40 ± .04 | .32 – .50 | 3.94 (.10) |
| Wood | .20 ± .02 | .14 – .26 | 9.37 (.24) |

**Table 3.** Ball roll data for eight different surfaces. Data are mean ± SD.

| Surface | Ball Speed (mph) | Cut/Spin (rpm) | Initial Ball Roll (rpm) | Forward Roll (inch) | Distance to true roll(") | Launch Angle (°) |
|---|---|---|---|---|---|---|
| **Robot** | | | | | | |
| Putt mat | 3.00 ± .12 | 9.12 ± 5.31 | 11.16 ± 14.70 | .41 ± .10 | 7.03 ± .47 | 2.00 ± .34 |
| Rubber | 3.00 ± .14 | 9.00 ± 5.55 | 1.13 ± 14.17 | .31 ± .11 | 6.00 ± .50 | 2.00 ± .27 |
| MFC | 3.06 ±.12 | 8.67 ± 5.48 | 11.78 ± 16.09 | 1.01 ± .39 | 16.77 ± 2.47 | 2.15 ± .31 |
| Carpet | 3.02 ±.10 | 9.53 ± 6.62 | 8.27 ± 16.49 | .70 ± .25 | 10.33 ± .87 | 2.06 ± .30 |
| **Human** | | | | | | |
| Brass | 2.98 ± .20 | 9.57 ± 4.91 | 25.55 ± 20.39 | 3.01 ± .17 | 5.97 ± .51 | .61 ± .71 |
| PTFE | 2.94 ± .29 | 9.50 ± 5.90 | -2.29 ± 13.77 | .93 ± .25 | 5.12 ± .10 | 1.08 ± .80 |
| Rubber | 2.85 ± .19 | 9.33 ± 5.38 | 15.87 ± 15.50 | .70 ± .20 | 3.94 ± .49 | 1.38 ± .70 |
| Wood | 3.04 ± .29 | 8.13 ± 5.66 | 11.52 ±10.28 | .54 ± .46 | 9.37 ± 1.85 | 1.98 ± .67 |

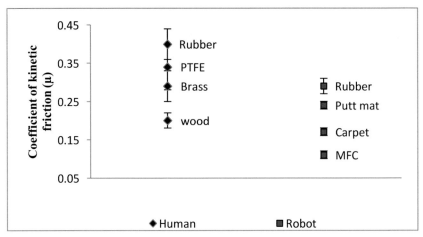

**Fig. 4.** Coefficient of kinetic friction (μ) derived from eight different surfaces using the two different methods.

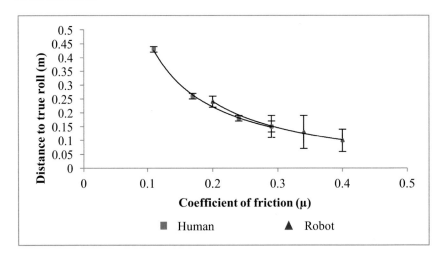

**Fig. 5.** Effect of coefficient of kinetic friction (μ) on distance to true roll for the two different methods.

## 4    Discussion

Considering the vast amount of golfing literature available, particularly within the area of putting, it's of some surprise that so little attention has been paid to the putting surface, in fact a significant emphasis has been placed on the effect of the golfer's equipment [14,15,16,17]. Whilst this has its significance within golfing research, this emphasis may be at the cost of other factors such as the putting surface, green reading and pace control. Therefore by paying attention to this area and more importantly the effect of the coefficient of friction of the putting surface and golf ball, it provides a significant addition to golfing literature which then has its opportunity to educate coaches, golfing manufacturers and ultimately the putting performers of the players themselves.

Table 2 shows that the rubber material that was used for the robot method and the brass material used for the human method had the same amount of kinetic friction (0.29), and both allowed the ball to enter true roll at the same distance (0.15m). This suggests that both methods are directly comparable.

Due to the large amount of time players spend traveling to and from competitions around the world, preparation can be detrimentally affected. Currently, the playing conditions of golf greens are measured and quantified through the use of a stimpmeter. In critique of this practice, the stimpmeter does not reflect the rolling characteristics of a golf putt. As highlighted in this study, a skidding ball has a significant bearing on pace control. The use of a stimpmeter eradicates this aspect of a balls motion. The use of the 'Quintic Ball Roll' system as a measuring tool may be used by coaches and greenkeepers to replicate putting green characteristics away from the golf course for players to familiarise themselves with the conditions in order to prepare themselves for competition play. Based on these suggestions it would be of interest for future research to investigate the relationship artificial surfaces, such as those used in this study, have with grass (outdoor golf greens). Manufacturing a number of artificial putting surfaces with diverse ranges of surface frictions would allow golfers and coaches to adapt their game, thus improving putting knowledge and performance.

# 5   Limitations

The parameters for the ball (launch angle, cut and speed) had some variance which may have affected the results. The surfaces used for the putting robot and the human subject differed which reduces the validity of the results.

# 6   Conclusion

It can be concluded that the amount of coefficient of friction that a surface holds has a major effect on the roll characteristics of a golf ball, with higher amounts causing the ball to enter true roll at an earlier stage when compared to surfaces with lower amounts. This has potential implications on the practice regimes of coaches and players' alike, thus directing attention to otherwise overlooked aspects of their practice. It could also be said to be of interest to course designers and green keepers when wishing to alter the difficulty of the course. Overall this study has explored a largely unexplored area. This study has the potential to be further developed and the effect of putting greens on ball roll characteristics can now be explored.

# 7    References

1. **Pelz, D.:** Dave Pelz's Putting Bible: 'The complete guide to mastering the green'. New York: Random House Digital (2000)
2.  Swash, H.: 'Championship Putting with Harold Swash' Southport: Yes! Golf Ltd (2001)
3.  Karlsen, J., Smith, G., & Nilsson, J.: The stroke has only a minor influence on direction consistency in golf putting among elite players. Journal of Sports Sciences, 26(3), .243-250 (2008)
4.  Pope, J., James, D., Wood, P., & Henrikson, E.: The effect of skid distance on dis tance control in golf putting. Procedia Engineering, 72, 642-647 (2014)
5.  Carr, G.: Sport Mechanics for Coaches. 2nd ed. Champaign, IL: Human Kinetics. 53-54 (2004)
6.  Palmer, G.: Physics for Game Programmers. United States of America: Apress.187 (2005)
7.  Epperson, B.P & Gadsden, E.T.: Golf. Birmingham, Ala. Pat. 5,383,664 (1993)
8.  Flom, D.G., & Beuche, A.M.: Theory of rolling friction for spheres. Journal of Applied Physics, 10, pp.1725 -1730 (1959)
9.  Hubbard, M., & Alaways.: Mechanical Interaction of the Golf Ball with Putting Greens. Science and Golf Ill: Proceedings of the World Scientific Congress of Golf, 54, pp.429 – 439 (1999)
10. Laws, W.G.: Optimal golf swing kinetics and kinematics. Science and Golf I. Proceedings of the First World Scientific Congress of Golf; pp.3-13 (1990)
11. Weber, A.: Green speed physics. USGA Green section record, 35(2), 12-15 (1997)
12. Hurrion, P.: Quintic Ball Roll Software (applications). [ONLINE] Available at: http://www.quinticballroll.com/Quintic_Ball_Roll_Contact_us.html (2103)
13. Normani, F.: The Physics of Bowling. Real-world-physics-problems.com. Available from: http://www.real-world-physics-problems.com/physics-of-bowling.html (Accessed April 20th 2014) (2009)
14. Brouillette, M., & Valade, G.: The effect of putter face grooves on the incipient rolling motion of a golf ball. Science and Golf V: Proceedings of the World Scientific Congress of Golf (2008)
15. Hurrion. P.D., & Hurrion, R.D.: An investigation into the effect of the roll of a golf ball using the c-groove putter. Science and Golf IV: Proceedings of the World Scientific Congress of Golf, 47, pp.531-538 (2002)
16. Brouillette, M.: Putter features that influence the rolling motion of a golf ball. Procedia Engineering, 2(2), pp.3223-3229 (2010)
17. Lindsay, N.M.: Sports Engineering, 6, pp.81-93 (2003)

# Modelling the Tactical Difficulty of Passes in Soccer

Michael Stöckl[1], Dinis Cruz[2] and Ricardo Duarte[2]

[1] University of Vienna, Vienna, Austria
[2] Universidade de Lisboa, Lisbon, Portugal

**Abstract.** A pass is an important action in soccer. Amongst other things, it allows moving the ball across the pitch quickly to create scoring opportunities and score goals. Passing performances are usually analyzed using descriptive measures such as number of passes or passing accuracy, however, neglecting the game context. This paper provides an approach to assess passes with respect to the tactical behavior of the participating teams. Results illustrate the application of the model by analyzing the passes played by players of different playing positions. Depending on the game context the players on different playing positions face different environments and, therefore, play passes of different tactical difficulty.

## 1 Introduction

The result of a soccer match is determined by the scored goals. The more goals a team scores the likelier it will win. Thus, scoring goals is a prominent aim in soccer and, likewise, preventing the opponent from scoring which increases the chance for winning as well. To score goals the ball needs to be moved to areas on the pitch where the probability for goal shots is high, such as the box or close to it. This makes ball moving actions to key factors in soccer. Generally, there are two ways to move the ball across a soccer pitch - dribbling and passing.

In the past there were several studies investigating passing performances in soccer. Some of them considered passing rates which describe the percentage of successful passes with respect to all passes a player/team played during a match [1, 4–6, 8]. Other studies counted passes [3] which belong to pass sequences leading to goals or investigated the length of passes resulting in goals [9].

However, most of those studies do not further distinguish between different kinds of passes, except [9] where the authors consider one pass characteristic at least. Often players are compared to each other using passing rates. A high passing accuracy is usually interpreted as good performance. On the one hand, when a player has a high passing rate he or she performed nearly as good as possible. On the other hand, a player showing a high passing accuracy might have played a lot of easy passes under no pressure to an area where the recipient is not pressed by the opponent either.

Therefore, the aim of this study was to develop a measurement which allows assessing passing in context of the game. The authors provide an approach to describe the tactical difficulty of passes based on the tactical behavior of the teams.

© Springer International Publishing Switzerland 2016
P. Chung et al. (eds.), *Proceedings of the 10th International Symposium on Computer Science in Sports (ISCSS)*, Advances in Intelligent Systems and Computing 392, DOI 10.1007/978-3-319-24560-7_17

## 2   Methods

**Materials** Data collected by Opta SportDaten AG from five English Premiere League matches from the season 2012/13 was used. For the difficulty calculations passes were described by triplets $(x, y, z)$, where $(x, y)$ is the positional information where a pass started or where it ended and $z$ is the information on the pass outcome. The outcome of a successful pass was denoted as 1, otherwise 0.

**Approach** Our approach focuses on the event pass itself in soccer. When a pass occurs during a match it is comprised by two actions: First, one player needs to play the ball somewhere aiming a teammate. Second, another player needs to receive it or at least a recipient needs to be recognizable unless it is a misplaced pass due to a misunderstanding or a technical failure. Therefore, if a pass is successful depends on two distinct actions. The player can do his or her best, but the pass can still be unsuccessful if the recipient performs badly. On the other hand, an actually badly played pass can still be successful if the recipient performs very well. However, the performances of the player as well as the recipient are not only constrained by themselves, but are also influenced by environmental and task constraints. Closer to the opponent's goal there will be put more pressure on the player and the recipient by the defending team than in the own half of the pitch when starting a play.

In this study we model the difficulty of a pass by relating it to the location of the player and the location of the recipient. The difficulty at a location on the soccer pitch is determined as the probability that a pass played from there or to there is successful. We argue, that on a professional level the players are able to play and receive passes in such a way that the pass is successful if both players were not disturbed at all. Thus, the interaction with the defending team is the most constraining source for a pass being successful. The more the player and/or the recipient are pressed the likelier a pass will be unsuccessful. Where and to which extent a defending team is pressing the opponent greatly depends on the respective team's tactic.

**Algorithm** The pass difficulty was modelled for players and recipients separately since they perform independently from each other and face different environments. The difficulty at the players' location and the recipients' location each were calculated as a continuous topology and for each team in each match separately. This ensures that the respective attacking and defending tactics of the participating teams are represented best. Based on the respective triplets $(x, y, z)$ the topologies were calculated using an algorithm provided by [7]. Since the details of the algorithm are described elsewhere only a short summary of the steps is provided below:

- A grid is put on the soccer pitch (grid size 2 m).

- Difficulty values are calculated at the grid nodes. Firstly, for each grid node all measured passing or receiving triplets are ordered in ascending order according to the distance to the respective grid node. Subsequently, the difficulty value at each grid node is calculated based on the z-values of the respective ordered data using an exponential smoothing (in this study the smoothing parameter $\alpha = 0.1$).
- To achieve a continuous topology of difficulty, the values between the grid nodes are interpolated based on the difficulty at the neighboring grid nodes. The interpolation was realized using a cubic smoothing spline interpolation [2] in order to remove rough edges in the resulting topology

$$\min_f \beta \sum_{i=1}^{n} \sum_{j=1}^{m} (z_{ij} - f(x,y))^2 + (1 - \beta) \iint (D^2 f(x,y))^2 dx dy \qquad (1)$$

where $D^2 = \frac{\partial^2}{\partial^2 x} + 2\frac{\partial^2}{\partial x \partial y} + \frac{\partial^2}{\partial^2 y}$ and $\beta$ is the smoothing parameter (in this study $\beta = 0.98$).

The tactical difficulty at any location can be extracted from the resulting topologies. According to the conditional probability concept the pass difficulty (PD) is defined as the product of the difficulty at the player's location (DP) and the difficulty at the recipient's location (DR)

$$PD = DP \cdot DR. \qquad (2)$$

DP and DR were weighted equally because both actions are equally important for a pass being successful. All calculations were conducted using MATLAB 2014b.

## 3    Results

Figure 1 illustrates the topology of the difficulty at the players' location (left) and the one of the difficulty at the recipients' location (right) of a team of an analyzed match. Values close to 1 (red areas) represent a high probability that passes from or to this location were successful and the smaller the values get (blue areas) the less passes were successfully played from or to this location. A soccer expert compared the topologies with the videos of the matches qualitatively. He corroborated that the topologies represent the difficulty on the pitch emerging from the teams' tactics.

As first application of the newly developed model we analyzed the difficulty of passes played by different groups of players (see table 1). The players were assigned to the groups goalkeeper (GK), defender (DF), midfielder (MF), or forward (FW) and the difficulty of played passes was analyzed with respect to all passes, successful passes, and unsuccessful passes.

Based on all passes a Kruskal-Wallis test showed a significant difference between the player groups (p=.000). Bonferroni corrected Mann-Whitney tests revealed that the passes played by GK significantly differ from passes played by

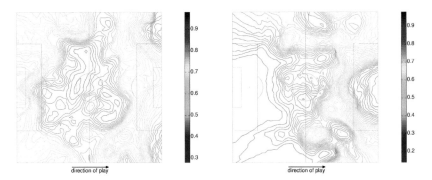

**Fig. 1.** Topology of the probabilities at the players' location (left) and a topology of the probabilities at the recipients' location (right). The probabibilty is color coded, blue marks smaller values and red marks higher values.

**Table 1.** Mean and standard deviation of PD are presented with respect to different player groups considering either all passes, successful passes or unsuccessful passes

|  | all passes | | successful passes | | unsuccessful passes | |
|---|---|---|---|---|---|---|
|  | mean | SD | mean | SD | mean | SD |
| **goalkeeper** | .525 | .206 | .634 | .173 | .369 | .137 |
| **defender** | .645 | .204 | .708 | .162 | .406 | .171 |
| **midfielder** | .606 | .206 | .671 | .163 | .376 | .175 |
| **forward** | .568 | .212 | .644 | .169 | .369 | .184 |

DF (p=.000), MF (p=.000), and FW (p=.004); further, the passes played by DF significantly differ from the ones played by MF (p=.000) and FW (p=.000); additionally, the passes played by MF significantly differ from those played by FW (p=.003).

With respect to successful passes a Kruskal-Wallis test showed a significant difference between the player groups as well (p=.000). Bonferroni corrected Mann-Whitney tests revealed that GK played significant different difficult passes than DF (p=.000) and MF (p=.000) and, furthermore, that DF played significant different difficult passes than MF (p=.000) and FW (p=.000).

Based on unsuccessful passes a Kruskal-Wallis test revealed a significant difference between the groups too (p=.017). However, Bonferroni corrected Mann-Whitney tests only showed a significant difference of played unsuccessful passes between GK and DF (p=.003).

Generally, the player groups statistically played passes of different difficulty. Therefore, only comparing those players with respect to passing accuracy is not decent since this performance measure does not take account of the game context. However, the differences between the player groups are smaller considering only successful passes or unsuccessful passes. In particular, GK and FW as well as MF and FW played passes of similar difficulty in those categories.

## 4 Conclusion

In this study an approach was developed to model the difficulty of passes in soccer with respect to the participating teams' tactics. The resulting indicator represents the 'tactical difficulty' of a pass considering the difficulty at the player's location and the difficulty at the recipient's location. This indicator can help shedding light on the passing performance of players which are usually analyzed based on passing accuracies.

## References

1. Bradley, P.S., Lago-Peñas, C., Rey, E., Gomez Diaz, A.: The effect of high and low percentage ball possession on physical and technical profiles in English FA Premier League soccer matches. Journal of Sports Sciences 31(12), 1261–70 (2013)
2. Fahrmeir, L., Kneib, T., Lang, S.: Regression. Springer, Berlin (2009)
3. Hughes, M., Franks, I.M.: Analysis of passing sequences, shots and goals in soccer. Journal of Sports Sciences 23(5), 509–14 (2005)
4. Lago-Peñas, C., Lago-Ballesteros, J., Rey, E.: Differences in performance indicators between winning and losing teams in the UEFA Champions League. Journal of Human Kinetics 27(1), 135–146 (2011)
5. Luhtanen, P., Belinskij, A., Hayrinen, M., Vanttinen, T.: A comparative tournament analysis between the EURO 1996 and 2000 in soccer. International Journal of Performance Analysis in Sport 1(1), 74–82 (2001)
6. Redwood-Brown, A.: Passing patterns before and after goal scoring in FA Premier League Soccer. International Journal of Performance Analysis in Sport 8(3), 172–182 (2008)
7. Stöckl, M., Lames, M.: Creating a Continuous Topography of Performance from Discrete Sports Actions. 7th Vienna International Conference on Mathematical Modelling, February, 2012. Mathematical Modelling 7(1), 814–818 (2013)
8. Wallace, J.L., Norton, K.I.: Evolution of World Cup soccer final games 1966-2010: Game structure, speed and play patterns. Journal of Science and Medicine in Sport 17(2), 223–228 (2014)
9. Yiannakos, A., Armatas, V.: Evaluation of the goal scoring patterns in European Championship in Portugal 2004. International Journal of Performance Analysis in Sport 6(1), 178–188 (2006)

# Convergence and Divergence of Performances Across the Athletic Events for Men and Women: A Cross-Sectional Study 1960 – 2012

Ian Heazlewood [1], Joe Walsh [1]

[1]Charles Darwin University

**Abstract.** There is a perception that performances in athletic events, such as the sprints, throws distance and hurdles events are converging for the top athletes in each event where access to quality training methods and performances at international competitions are readily available for international ranked athletes. Some researchers cite biological reasons that may explain the suggested convergent tends in athletics. The research hypothesis and aim was to evaluate if performance convergence was occurring using the method of regression analysis of the mean of standard deviations, an index of performance dispersion or divergence, for the top 20 male and female athletes across the majority of athletic events from years 1960 to 2012. The results indicated that most male events have reached stasis, neither converging nor diverging whereas female events are currently undergoing performance convergence. Factors such as access to high performance coaches, sports scientist, relevant sports nutritionists and quality medical support are producing the level playing field that is now so level that minimal performance diversity exists between the top 20 male athletes across many events. The reverse situation applies to female athletes who are still displaying performance convergence and who are progressing towards a level playing field.

## 1 Introduction

There is a perception that performances in athletic events, such as the sprints, throws and hurdles events are converging for the top athletes in each event. That is, the top athlete performances over the years are getting similar or homogeneous as a result of top ranked athletes utilising similar training methods, where access to training methods and performances at international competitions are readily available for international ranked athletes via You Tube or high performance coaching videos. This enables other athletes to readily access skills and techniques for running, hurdling, throwing and jumping. There is now an opportunity for high performance athletes to compete frequently against one another on the international athletic circuit via annual Diamond League and other International Federation of Athletic Associations (IAAF) permit competitions that frequently serve as preparatory and lead-up competitions for World Championships held every two years, the Olympic Games every four years and the Commonwealth Games every four years. High performance athletes now can gain

© Springer International Publishing Switzerland 2016
P. Chung et al. (eds.), *Proceedings of the 10th International Symposium on Computer Science in Sports (ISCSS)*, Advances in Intelligent Systems and Computing 392, DOI 10.1007/978-3-319-24560-7_18

145

significant financial rewards in terms of prize money, appearance money and sponsorships providing access to high performance coaches, sports scientists, relevant sports nutritionists and quality medical support. The suggestion is the level playing field is now so level that minimal performance diversity exists between athletes. In many 100m sprint races only hundreds of seconds separate first to sixth.

A number of quotations highlight this conceptual model where Weigel [1] a national coach of German race walkers states, "The top of the world has converged. " Pfützner [2] states, "Density of performances will further increase and the battle for medals will become harder." Finally, Grabner [3], states, "In this context (after some years of women's pole vault) performance level and density, as well as the dissemination of the women's pole vault to many different countries has increased. The demand for knowledge from sports science in this still young discipline is permanently increasing and the belief is there will be an increase in performance density or performance convergence/homogeneity within the top ranked athletes' across all athletic events.

Truyens, De Bosscher and Heynels [4] have suggested a possible mechanism for this increased performance density or performance convergence where they state, "An increasing number of countries have developed strategic approaches in their pursuit of success and there is an international trend towards a homogenous model of elite sports development." Does the homogenous model result in homogenous training practices across different events that eventually produce homogenous athletes at the elite level?

Other researchers cite biological reasons that may explain the suggested convergent trends in athletics. In a conceptual paper "How fast can a human run," Richmond [5] hypothesized that human biological or evolutionary limits are the reasons for performance convergence as metabolic and exercise physiological factors and thermodynamic principles set definable limits and has set the limit in the 100m to 9.27s. This might be the situation in the men's 100m where the average of the top seven 100m men's finalist times at the 2012 London Olympics was 9.824s, so the suggestion is 100m athletes may have converged to this biological theoretical limit [6].

The past and current literature does not adequately explain the conceptualisation or present empirical evidence supporting or refuting such conceptualisations as the quantitative models have not been derived. Evaluating such a trend requires the evaluation of data over a significant timeframe, such as decades and assessing indices of statistical homogeneity, such as a reduction in performance dispersion. One index in this context would be reflected in reduction in variances and standard deviations for the top athletes. It is important to highlight the convergence of athletic performance does indicate overall performance improvement in an event, as Heazlewood and Walsh [7, 8] indicate that in men's and women's throwing events overall performances have displayed performance declines from the late 1980's to present. The research hypothesis and aim was to evaluate if performance convergence was occurring there should be a reduction of performance dispersion as indicated by reducing standard deviations of the top 20 performances as rated by the International Association of Athletic Federations (IAAF).

# 2    Methods

## 2.1    Sample

The sample in the study consisted of the 20 top IAAF ranked athletes in the following events for men and women senior athletes based on legal outdoor performances [9]. Events were measured to the nearest .01 seconds (automatic timing) for sprint, hurdles and running events and to the nearest centimetre for throwing and jumping events. The men's events were 100m, 200m, 400m, 800m, 1500m, 5000m, 10000m, 110m hurdles, 400m hurdles, 3000m steeple chase, long jump, high jump, pole vault, shot put, javelin, discus and hammer. Data were from IAAF competition years 1960 to 2012 and n of cases was 18,020 male athletes. The women's events were 100m, 200m, 400m, 800m, 1500m, 110m hurdles, 3000m, long jump, high jump, shot put, javelin, discus and hammer. Newer athletic events for women, except the hammer, were not included as the number of years of competition was minimal. Once again, the data were from IAAF competition years 1960 to 2014 and n of cases was 13,520 male athletes. It is important to note that fully automatic timing was available in 1960 by the Omega Time Recorder and clocks were added to slit cameras for automatic time-stamping and were accurate to the 100th of a second.

## 2.2    Analytical Methods

The performance for the top 20 athletes for each event mentioned were inputted into a SPSS Statistics Version 22 [10] data matrix and the means, ranges, minimum and maximum scores, variances and standard deviations were calculated for each event. Mean scores represented a measure of centrality and variances and standard deviations as a measure of dispersion of scores. This standard deviation for each year can be regressed with year of performance to evaluate trend.  Regressing the standard deviations on time with year of competition would indicate a negative slope and increasing convergence and the alternative solutions would apply for increasing divergence that of positive slope and no significant change would produce a zero or horizontal line. The regression solution was bivariate linear regression. Other non-linear regression models were also applied. The solutions were derived for each event over the 1960 to 2014 timeframe to evaluate performance trends.

The intention of linear regression analysis to; develop an equation that summarizes the relationship between a dependent variable and an individual or set of independent variables or variable; identify the subset of independent variables or variable that are most useful for predicting the dependent variable; and finally to predict values for a dependent variable from the values of the independent variable(s) [11]. Event specific models were derived based on gender specific events. Regression fit criteria include the amount of explained variance or $R^2$, levels of significance, the fit between the actual and predicted values and evaluating error or the residuals the difference between the actual and predicted scores. It is to be emphasised the overall objective was not to assess if the events were displaying increases or decreases in event performance over time, such as changes in world records. The focus was on whether or not the performances of the top 20 athletes were getting closer or further apart from each other over time or to assess if athlete performances are becoming more homogenous in nature.

## 3    Results

The results for bivariate linear regression indicated some very different outcomes for different genders and these are presented in table 1 for males and table 2 for females. The men's results indicate that only four of the seventeen events studied displayed convergent or divergent trends. Specifically, 100m, 200m and 3000m steeple chase are diverging, whereas 110m hurdles is converging. What is important to highlight is all other events displayed non-significant regression slopes. This indicates that many events are already static in terms of event convergence, whereas the 110m hurdles is currently undergoing further convergence.

The female trends are significantly different from the male trends as majority of events evaluated are undergoing convergence. Specifically, 200m, 400m, 100m hurdles, 800m, 1500m, 3000m, shot put and discus. Only the 100m, long jump high jump and javelin displayed non-significant slopes and may indicate that convergence is now static in these events. The hammer is a new event and was included to assess the trends in a new event as it was first introduced into the Olympic Games in Sydney 2000.

**Table 1.** Men's athletic events, mean of standard deviation, standard deviation- standard deviation, standardised beta and p-values. ns represent non-significant relationship for beta values.

| Event | Mean s.d. | s.d.-s.d. | Beta | p |
|---|---|---|---|---|
| 100m | .070 | .018 | .393 diverge | .004 |
| 200m | .165 | .055 | .377 diverge | .005 |
| 400m | .316 | .088 | -.134 ns | .337 |
| 110m hurdles | .136 | .026 | -.443 converge | <.001 |
| 400m hurdles | .486 | .117 | -.254 ns | .070 |
| 800m | .655 | .173 | .050 ns | .724 |
| 1500m | 1.409 | .361 | -.035 ns | .802 |
| 3000m steeple | 4.287 | 1.223 | .571 diverge | <.001 |
| 5000m | 4.793 | 1.442 | .013 ns | .928 |
| 10000m | 12.962 | 4.326 | .173 ns | .215 |
| Long jump | .128 | .033 | -.253 ns | .068 |
| High jump | .026 | .008 | -.251 ns | .070 |
| Pole vault | .079 | .016 | -.159 ns | .257 |
| Shot put | .508 | .093 | -.006 ns | .965 |
| Javelin | 2.421 | .584 | -.114 ns | .419 |
| Discus | 1.642 | .303 | -.049 ns | .728 |
| Hammer | 1.644 | .395 | -.205 ns | .149 |

**Table 2.** Women's athletic events, mean of standard deviation, standard deviation- standard deviation, standardised beta and p-values. ns represent non-significant relationship for beta values.

| Event | Mean s.d. | s.d.-s.d. | Beta | p |
|---|---|---|---|---|
| 100m | .109 | .022 | -.164 ns | .240 |
| 200m | .248 | .054 | -493 converge | <.001 |
| 400m | .592 | .159 | -.476 converge | <.001 |
| 100m hurdles | .161 | .048 | -.842 converge | <.001 |
| 800m | .853 | .658 | -.471 converge | <.001 |
| 1500m | 2.233 | .731 | -.505 converge | <.001 |
| 3000m | 5.838 | 2.133 | -.454 converge | .004 |
| Long jump | .125 | .039 | -.017 ns | .901 |
| High jump | .031 | .007 | -.118 ns | .400 |
| Shot put | .749 | .174 | -.433 converge | .001 |
| Javelin | 2421 | .584 | -.114 ns | .419 |
| Discus | 2.102 | .494 | -.361 converge | .008 |
| Hammer | 1.644 | .396 | -.205 ns | .149 |

An example of the trend for event divergence for men's 100m sprint is presented in figure 1. An example of the trend for event divergence for women's 100m hurdles is presented in figure 2. The significant positive slope or divergence and significant negative slope or convergence can be observed in figure 1 and figure 2 for the men's 100m and women's 100m hurdles respectively.

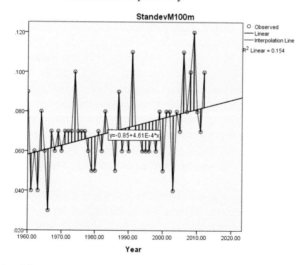

**Figure 1.** Trend for event divergence for men's 100m sprint based on linear regression equation. (Y-axis is the s.d. of top 20 performances.)

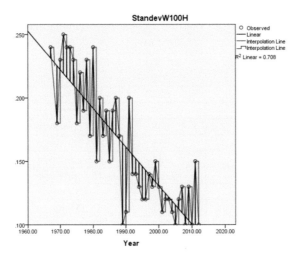

**Figure 2.** Trend for event convergence for women's 100m hurdles based on linear regression equation. (Y-axis is the s.d. of top 20 performances.)

## 4    Conclusion

From table 1 it can be observed that the male performances for the top 20 ranked athletes from 1960 to 2012 across the majority of events display neither convergence nor divergence and at this point in time display stasis. This does not mean that the top athletes are not changing in terms of mean performance just that the variance between athletes has been relatively constant over the last 52 years. In the men's 100m and 200m where divergence has been noted two athletes Bolt and Blake from Jamaica have displayed significantly higher performances than rival athletes including recent world records in the 100m and 200m, which has increased the performance dispersion (variance) or performance divergence. This result is inconsistent with the predictions of Richmond [5] for the 100m; however the other running, throwing and jumping events may have reached the limits of convergence and reflects that males in terms of training time, training resources, nutrition, medical support, training adaptability and competitive opportunities are becoming more homogenous in high performance outcomes.

The female trends are very dissimilar to the males as the majority of their events display performance convergence. That is the performances for the top 20 female athletes in athletics across many events are being pushed closer together, however at this point in time based on the 54 year trends, they may have not reached the limits of reduced performance variability. These trends may indicate the factors identified as promoting stasis in performance variability in males are not as developed in females, such as top ranked males athletes utilising similar training methods, now are full-time paid professional athletes and having sufficient time for optimal training. Once again, it must be emphasised that for female athletes, although performance convergence is occurring, that is the actual performance differences between athletes is being reduced, paradoxically it is important to note that in many women's events the mean of the top 20 performances is actually declining. Further monitoring of performance trends both in mean performances of the top 20 athletes, as well as performance variances will indicate whether or not humans have reached the limited of human performance cross the different athletic events that represent such constructs as human speed, strength, power and endurance.

# References

1. Weigel, R.: National Coach of German Race Walkers: "Die Weltspitze ist näher zusammengerückt"https://www.leichtathletik.de/news/news/detail/leichtathletikde-analyse-gehen-frauen-1/ (2014)
2. Pfützner, A.: Internationale und nationale Tendenzen der Leistungssportentwicklung auf der Grundlage der Ergebnisse der Olympischen Spiele in Peking 2008. Leistungssport, 39 (1), 17-20. (2009)
3. Grabner, S.: Energetische und konditionelle Aspekte beim Stabhochsprung der Frauen. Leistungssport, 34 (5), 33-38. (2004)
4. Truyens, J., De Bosscher, V., Heynels, B.: The SPLISS Athletics Project: a Resource-based of Elite Athletics Policies. New Studies in Athletics, 27, 133 - 139 (2012)
5. Richmond, J.: How Fast Can a Human Run? New Studies in Athletics. 27:4; 57-62, (2012)
6. Dick, F.: 2012 Olympic Performance Assessment – A European Perspective. New Studies in Athletics. 27:4; 67-74, (2012)
7. Heazlewood, I., Walsh, J.: Mathematical Models that Predict Athletic Performance in the Men's Throwing Events at the Olympic Games. Proceedings of the 12th Australasian Conference on Mathematics and Computers in Sport. Bedford, A. & Heazlewood, T. Eds. June 25th - 27th, 2014, Darwin, Northern Territory, Australia. MathSport (ANZIAM). pp. 48-53. (2014)
8. Heazlewood, I., Walsh, J.: Mathematical Models that Predict Athletic Performance in the Women's Throwing Events at the Olympic Games. Proceedings of the 12th Australasian Conference on Mathematics and Computers in Sport. Bedford, A. & Heazlewood, T. Eds. June 25th - 27th, 2014, Darwin, Northern Territory, Australia. MathSport (ANZIAM). pp. 54-60. (2014) http://www.iaaf.org/records/toplists/sprints/100-metres/outdoor/men/senior/ (2015)
9. International Amateur Athletic Federation. Top Lists. http://www.iaaf.org/records/toplists/sprints/100-metres/outdoor/men/senior/ (2015)
10. SPSS INC. IBM SPSS Statistics Version 22 Statistical Software: Core System Users' Guide. SPSS Inc. (2014)
11. Hair, J., Block, W., Babin, B., Anderson, R., Tatham, R.: Multivariate Data Analysis. (6th Ed.). Pearson - Prentice Hall. (2006)

# Introduction of the relative activity index: Towards a fair method to score school children's activity using smartphones

Emanuel Preuschl, Martin Tampier, Tobias Schermer, and Arnold Baca

University of Vienna, Institute of Sports Science,
Auf der Schmelz 6a, A-1150 Vienna, Austria

**Abstract.** The on-going technical progress in pervasive computing influences our daily activities more and more. In sports, smartphone apps are used to support athletes during training - either by giving plain physiological or motivating feedback. The Mobile Motion Advisor (MMA) is such kind of system, developed for the use in school children's physical education. A school class resembles a highly inhomogeneous group regarding their physical fitness. Hence, considering absolute performance parameters might not be an appropriate way to grade the school children's efforts. With the introduction of the relative Activity Index ($rAI$), we propose a method that approaches fair grading based on the technologies of the MMA. The $rAI$ is a calculated value that resembles the relative activity of an individual within a group, based on the number of steps of each individual. As we found out, $rAI$ correlates significantly ($r = 0.620, p < 0.01$) with the $\dot{V}O_2max$ which allows to deduce an estimated individual $rAI_{est_m}$. Comparing the estimated $rAI_{est}$ with the actual $rAI_m$ may enable us to grade efforts within the context of the games dynamics and the individual fitness.

## 1 Introduction

Information and communication technologies gradually permeate our lives. Technical or informational aid is given by smart phones and other portable devices. Hikers and runners more and more rely on topographic and physiologic data provided by such devices and appropriate sensors, respectively. In an attempt to investigate if this trend could be used for activity promotion amongst school children, a mobile coaching system has been adapted to physical education classes - resulting in the Mobile Motion Advisor (MoMA) [1]. The students are equipped with a mobile device and commercially available wireless sensors in order to acquire sports specific parameters during exercising. The measured data is buffered locally and forwarded via the Internet to a server. Based on the collected data, feedback instructions can be generated by experts and sent back to the exercising person. In addition, sub-modules may be integrated into the server application thus implementing algorithms for processing the acquired data. One sub-module, a virtual game named 'Beat-my-Activity (BMA) was developed to run concurrently to small sided sports games.

© Springer International Publishing Switzerland 2016
P. Chung et al. (eds.), *Proceedings of the 10th International Symposium on Computer Science in Sports (ISCSS)*, Advances in Intelligent Systems and Computing 392, DOI 10.1007/978-3-319-24560-7_19

In comparison to sport teams, school classes are likely to be more heterogeneous in their member's attitudes towards physical activity, their interests in sports and their performance levels. The idea behind BMA is giving an incentive to physical exertion to those students who may not have the abilities to compete with their peers due to their fitness or skills. BMA is designed to give every student the chance to win by rating his/her performance in relation to his/her aerobic capacity, then classifying the rating in three groups - leaders, pursuers and laggards and finally giving feedback on that status in the game periodically. Although maximal oxygen saturation ($\dot{V}O_2max$) is not the sole determinant of athletic performance potential, it's considered that the $\dot{V}O_2max$ is a good predictor of athletic performance when a heterogeneous group of athletes with quite different athletic abilities is studied [6]. Step counts have been used to quantify levels of activity of school children in several studies as Holfelder and Schott report in their review [5]. BMA uses commercially available wireless stride sensor's to count the steps during a session ($N_m$).

In team sports, the athletes may not play for the full duration of a game. Therefore, the BMA normalizes the effort of a student by the time he/she participated in the game which we call the activity index ($AI$). As not only the individual motivation of an athlete but also the dynamics of the game affect the intensity of exertion, AI is related to the group's mean AI. This relation is called relative activity index ($rAI$). With a $rAI$ around 1, the athlete is as active as the mean of the group, taking into account the dynamics of the game. Using an individual's $\dot{V}O_2max$-level, could the BMA set an appropriate $rAI$ level to judge whether the corresponding athlete is over- or under-performing? Aim of this study was to find out, if $\dot{V}O_2max$ and $rAI$ can be used together to build a model for fair rating of individual effort in sports games in school.

## 2   Methods

### 2.1   Participants

A group of students ($n = 25$, 11 male, 14 female, age $= 15.6 \pm 0.84$ years, body mass $= 70.8 \pm 17.23$ kg) attending a high school for economics participated in this study. They signed informed consent and the local ethics committee approved the project.

### 2.2   Procedure

First, each participant's age was recorded and body mass was determined using a personal scale. Then the students performed a 1-mile jog test for estimation of $\dot{V}O_2max$ [4]. For the remainder of the semester, the students participated in their weekly scheduled physical education lessons. A team sport game is always part of the lesson and takes between 20 to 50 minutes. The participants fasten the sensors and the smart phone and log in to the BMA app before the game is started. According to the teacher's instructions the games are played gender mixed or separated. The stride data was used to determine the relation of $\dot{V}O_2max$ and $rAI$. The students received feedback on their step count and actual heart rate during their activities.

## 2.3 Data acquisition and processing

The protocol of the 1-mile track jog was administrated using a heart rate sensor (Soft Strap Heart Rate Monitor, Garmin Ltd) and a stride sensor (Foot Pod, Garmin Ltd). These sensors transmit the data to a smart phone running the MoMA-App module for the 1-mile track jog, which supervises running speed and heart rate, covered distance and end time. The MoMA 1-mile track jog module gives feedback to the user about the covered distance every 100 m. At the end of the bout it gives feedback on the estimated $\dot{V}O_2max$. Concurrently, all data are stored and can be accessed via the MoMA webserver. In Baca et al. [1] more technical details on the MoMA are presented. BMA records the activity of each individual by counting the number of steps ($N_m$) via a commercially available stride sensor continuously during the game. $N_m$ is then normalized by the time the student is participating in the BMA game ($t_m$) resulting in the activity index ($AI$). By relating $AI$ with the arithmetic mean $AI$ of the group we calculate the relative activity index ($rAI$). By this, the games dynamics are taken into account and individual performance of different sessions may be compared. $rAI_m$ for a subject $m$ is determined with equation 1.

$$rAI_m = \frac{\frac{N_m}{t_m}}{\frac{1}{n}\sum_{i=1}^{n}\frac{N_i}{t_i}} \qquad \text{where } 1 \leq m \leq n \qquad (1)$$

## 2.4 Statistics

After the collection of each student's data over a semester the individuals' mean $rAI$ was determined. Statistical analyses were performed using SPSS ver. 22 (SPSS Inc., Chicago, Illinois). $\dot{V}O_2max$ and $rAI$ were tested for normal distribution using a Shapiro-Wilk Test and descriptive statistics were performed. The relations between $\dot{V}O_2max$ and $rAI$ were determined using Pearson's correlation coefficient if the data didn't differ significantly from normal distribution. Significance level was set to $p = 0.05$, for high significance to $p = 0.01$. Effect size was determined by the Pearson's correlation coefficient and is considered to be of practical relevance above $r = 0.2$, of moderate effect for $r = 0.5$ and strong for $r = 0.8$ and above [2,3].

## 3 Results

Descriptive statistics of $\dot{V}O_2max$ and $rAI$ are shown in Table 1. The Shapiro-Wilk tests indicate that $\dot{V}O_2max$ and $rAI$ do not differ significantly from normal distribution. We found a highly significant Pearson correlation of moderate effect between the students' $rAI$ and their $\dot{V}O_2max$ ($r = 0.620, p = 0.001$).

**Table 1.** Descriptive statistics of $\dot{V}O_2max$ and mean $rAI$ (Pearson correlation $r = 0.620, p = 0.001$)

| variable | | min | max | mean | sd |
|---|---|---|---|---|---|
| $\dot{V}O_2max$ $[ml\,kg^{-1}\,min^{-1}]$ | | 28.74 | 55.68 | 40.33 | 6.79 |
| $rAI$ | [-] | 0.40 | 1.82 | 0.98 | 0.31 |

## 4  Conclusion

As correlation between $\dot{V}O_2max$ and $rAI$ is highly significant and of medium to strong effect, we propose to estimate a student's expected $rAI$ ($rAI_{est}$) using a linear regression equation.

$$rAI_{est_m} = a + b\,\dot{V}O_2max_m \qquad (2)$$

The coefficients $a$ and $b$ in equation 2 are dependent on the actual game's dynamics. This affects the calculation of $rAI_{est_m}$ and helps to deal with games characterized by less running activity like dodge ball. In these games, a student with lower $\dot{V}O_2max$ might be able to be as active as a fitter peer. For the rating of a subject $m$, the difference between $rAI_{est_m}$ and the actual $rAI_m$ ($\Delta rAI_m$) is determined.

$$\Delta rAI_m = rAI_m - rAI_{est_m} \qquad (3)$$

However, a student with a negative $\Delta rAI_m$ is less active than expected - a positive value indicates more activity. By providing appropriate feedback we want to motivate the students to keep pace or to top the peers with regard to the expected performances.

## References

1. Baca, A., Kornfeind, P., Preuschl, E., Bichler, S., Tampier, M., Novatchkov, H.: A server-based mobile coaching system. Sensors 10, 10640–10662 (2010)
2. Cohen, J.: Statistical power analysis for the behavioral sciences. Lawrence Erlbaum Associates, 2 edn. (1988)
3. Ferguson, C.J.: An effect size primer: A guide for clinicians and researchers 40(5), 532–538 (2009)
4. George, J.D., Vehrs, P.R., Allsen, P.E., Fellingham, G.W., Fisher, G.: $\dot{V}o_{2max}$ estimation from a submaximal 1-mile track jog for fit college-age individuals 25(3), 401–106 (1993)
5. Holfelder, B., Schott, N.: Relationship of fundamental movement skills and physical activity in children and adolescents: A systematic review 15, 382–391 (2014)
6. Noakes, T.D.: Implications of exercise testing for prediction of athletic performance: a contemporarty perspective. Medicine and Science in Sports and Exercise 20, 319–330 (1988)

# Performance Analysis in Goalball

## Semiautomatic specific software tools

Christoph Weber & Daniel Link

Chair of Training Science and Sport Informatics – Technical University of Munich (TUM)

**Abstract.** Goalball is one of the paralympic disciplines designed for visually impaired athletes. Two teams, each consisting of three players try to score goals by rolling the ball into the oppositions' net. All players wear additional blindfolds to guarantee that players are equally impaired. The ball contains bells allowing players to echolocate its movement. Goalball-specific software was developed for performance analysis ("GoalScout", "GoalView" and "GoalTrack"). For practical and intuitive scouting purposes the software runs on tablet PCs. Additionally, positional data extracted form video sequences provides precise information. GoalTrack accurately measures the speed of the ball. The purposes of this project were to: a) design a software prototype that efficiently analyzes performance and, b) analyzes key aspects of Goalball. Bowling patterns were characterized for all teams and players. Significant scoring sectors were identified (Sector 3 and 7; $\chi^2 = 7.50$, $p \leq .006$). A correlation between ball speed and success rate is, especially in women games, statistically significant (Men: $r = .65$; Women: $r = .90$).

**Keywords:** Goalball, performance analysis, ball speed

## 1 Introduction

Goalball, a Paralympic sport since 1976 in Toronto, is designed for visually impaired athletes. The goal is that two opposing teams, each consisting of three players try to score goals by rolling the ball into the oppositions' net. Players wear blindfolds to guarantee equal impairments. The ball contains bells allowing players to echolocate movements. So far only very few studies in the field of Goalball performance analysis can be found. This project was funded by German Institute of Sport (IIA1-070405/12-13 and IIA1-070406/14) to develop state of the art software for performance analysis in Goalball.

## 2 Purpose

The purpose of this study was to light up performance in Goalball. Based on the model "Systematische Spielbeobachtung" by Lames [1], three software tools, which are specific to Goalball were developed. Grounded on the experience in Beachvolleyball

© Springer International Publishing Switzerland 2016
P. Chung et al. (eds.), *Proceedings of the 10th International Symposium on Computer Science in Sports (ISCSS)*, Advances in Intelligent Systems and Computing 392, DOI 10.1007/978-3-319-24560-7_20

157

[2] "GoalScout" is a data collection tool, while "GoalView" analyze data. "GoalTrack" accurately measures the speed of the ball by using several methods of computer vision to detect the ball. Study intends to provide information's about general structure of performance as well as the influence of target sector and ball speed.

## 3    Software Design and Development

For practical scouting purposes GoalScout runs on tablet PCs. Additionally, positional data extracted form video sequences provides precise information about start and goal sector. More frequently used game characteristics appear situation-related next to the video, which leads to a very intuitive and efficient way of scouting. Based on just 4 to 9 "clicks", up to 18 game characteristics can be derived for each bowl (see Table 1 and Figure 1). GoalView is a specific data analysis tool for Goalball. Using different filter settings, each video sequence of a bowl can be analyzed. Furthermore, reports (e.g. statistical, bowl-distribution, Performance Index etc.) can be created in real-time. To determent ball trajectory, GoalTrack uses a multistage approach of computer vision. Initially, by using Background Substraction as well as Color and Match shaping the position of the ball is detected from the video frame. Following a localization of ground contact points of the ball (minimum of pixel coordinates) plus a projection of ball position to ground level at these times. A result of this, is the regression line of the ball through floor contact points. According to Cohen, inter-rater-reliability between manual and automatically notation of sectors is high ($\kappa$=.807). Bland and Altman analysis [3] demonstrate a high level of agreement between GoalTrack and Utilius easy inspect. For the 95% confidence intervals a lower and higher limit of 1.3 m/s were detected.

**Table 1.** Game characteristics, captured for each bowl

| Game ID | Bowl ID |
|---|---|
| Timecode | Team/Player offense |
| Team/Player defense | Start position |
| Defense position | Ball direction |
| Angel of defense | Result of bowl |
| Used technique | Flight path of ball |
| Kind of defense | Penalty |
| Bowl after penalty | Bowl after change court sides |
| Bowl after scored goal | Bowl after the change of position of a player |

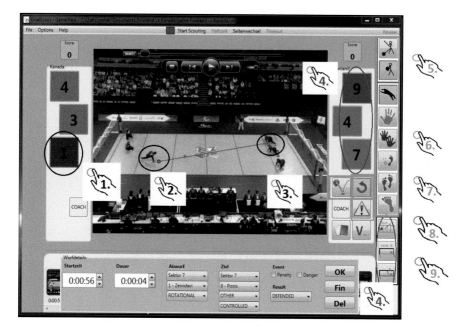

**Fig. 1.** User interface Goalscout. Up to 18 game characteristics can be derived by just 9 "clicks"

# 4    Performance Analysis in Goalball

## 4.1    Data Set

Altogether 113 Goalball matches were analyzed (Paralympics 2012, European Championships 2013, 63 men games, 50 women games). Software tools were used to analyze each bowl (n=21058) regarding to all 18 game characteristics listed in Figure 1. Furthermore GoalTrack was used to measure ball speed (n= 8397).

## 4.2    Results

No significant difference between male and female games in regards to no. of bowls per game were detected (n=60, Men = 187.95 ± 20.52; n= 53, Women = 184.60 ± 20.65). The scoring rate for males is higher than for females (n = 11828 Men = 5.44% ± 2.5; n = 9230 Women = 3.67% ± 1.5). Rotational bowling technique is used by male players more often (n = 11828 Men = 110.69 ± 42.80 bowls per game; n = 9230 Women = 45.54 ± 36.23 bowls per game). So called "bouncing ball" are used more often by male players as well (n = 11828 Men = 115.60 ± 40.12 bowls per game; n = 9230 Women = 48.54 ± 23.83 bowls per game). Scoring sector analyses show significance differences on sector 3 and 7 (n = 19754; $\chi2$ = 7.50, p≤.006). A correlation between scoring rate and ball speed (see Fig. 2; n= 8397; Men: r = .65; Women: r = .90) is statistically significant. KS-Test showed, that fast bowling is evenly distributed across matches (n = 899; D = .884, p<.01).

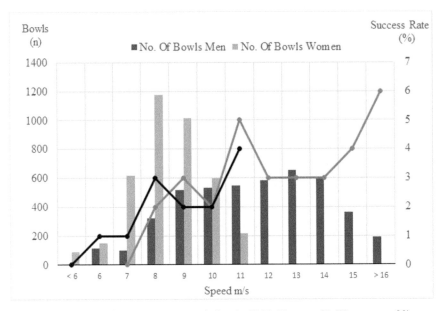

**Fig. 2.** Ball speed – success rate correlation (n=8397; Men: r= .65; Women: r = .90)

# 5    Conclusions

In this study, an investigation on performance indicators in Goalball was conducted. Scoring sector analysis proved coaches and players opinion that it is easier to score on sectors, where defensive responsibilities of two players overlaps. Significant correlation between scoring rate and ball speed, especially in women games, recommend a training of basic strength abilities. Additional research should further investigate underlying causes of observed bowling patterns as well as the impact rule changes.

# References

1. Lames, M. (1994): Systematische Spielbeobachtung. Münster: Philippka (Trainerbibliothek / Deutscher Sportbund, Bundesausschuss Leistungssport, Bd. 31).
2. Link, D. & Ahmann, J. (2013a). Spielanalyse im Beachvolleyball. Leistungssport, 43 (1), 58 - 63.
3. Bland, J.M. & Altman, D.G. (1986). Statistical methods for assessing agreement between two methods of clinical measurement. Lancet, i, 307– 310.

# Markov Simulation by Numerical Derivation in Table Tennis

Sebastian Wenninger and Martin Lames

Department of Training Science and Sport Informatics
Technische Universität München, Germany

**Abstract.** This study tries to assess the impact of different tactical behaviors on the winning probability in table tennis by mathematical simulation. Using discrete states of the game and transitions between them by means of the Markov chain model, the numerical derivation of winning probabilities is calculated under assumption of the Markov chain property. The method introduced in this study improves the traditional model by omitting the necessity to perform a second modeling step to define the difficulty of tactical behaviors. Application of our method on table-tennis shows meaningful results like the identification of long rallies as highly influencing stroke types and the effects of risky play on winning probability in different game situations.

## 1   Introduction

To assess performance indicators in sports which are hard to reproduce in reality, mathematical simulations are commonly used as auxiliary resources. The impact of a tactical behavior through means of Markov simulation is usually determined by the following steps: Under assumption of the Markov property, the observation system has to be modeled as a Markov chain. Then, transition matrices can be calculated for the individual games. In a second step, to analyze the impact of a behavior on the winning probability of a player, the transition matrices have to be altered in such a way, that the new transition probabilities represent the desired deviation in tactics. The resulting difference in winning probability gives an estimate of the importance of the modeled behavior. This deflection in tactical behavior requires an additional modeling step, because the simulation has to account for the relative difficulty of the behavioral change. Additionally, in order to retain the stochastic nature of the transition matrices, the tactical deviations have to be compensated by other transition probabilities, so that the transitions per row still add up to one. This original method was developed in tennis [3] and found applications in different sports, e.g. table tennis [2], water polo [4] and handball [5]. This study introduces an improvement of the method cited above. It skips the second step of modeling by taking the directional derivative of the function calculating the winning probabilities from the transitions. As there is no analytical derivative, numerical derivation is used. This improvement frees from the task of modeling alterations of game behavior of comparable difficulty.

P. Chung et al. (eds.), *Proceedings of the 10th International Symposium
on Computer Science in Sports (ISCSS)*, Advances in Intelligent Systems
and Computing 392, DOI 10.1007/978-3-319-24560-7_21

161

## 2    Methods

### 2.1    Design

A table tennis game was studied by means of a single observation system, which introduced the following 14 states to model a rally: first to fifth stroke, more than fifth stroke and point, each for player A and B. Hence, a single rally can be seen as a system that alternates between the stroke states of both players. The state systems commonly start with a serve (first stroke) and end with a point for either player A or B.

### 2.2    Participants

In this study, 259 matches of world-class male and female table tennis players were assessed. The examined games are spanning over the time of 4 years (2008-2012) and were played in official tournaments (ITTF, ETTU, IOC). The dataset contains the length (stroke numbers) and winner of all rallies, as well as the final game result for each game.

### 2.3    Simulation

To compute the derivative of the winning probability of a player for a particular stroke number, it can be regarded as a function $f(\hat{x}) = f(x_1, \cdots, x_{36})$ in the multi-dimensional space of the Markov model. This function takes the relevant transition probabilities from the Markov transition matrices as input parameters and returns the winning probabilities for both players in a two-dimensional vector. Since no analytical description of this function is available, numerical derivation has to be used.

#### Numerical Derivation
The derivative of a function is commonly defined as the forward difference

$$f' = \lim_{\hat{h} \to 0} \frac{f(\hat{x} + \hat{h}) - f(\hat{x})}{|\hat{h}|} \tag{1}$$

which is already in numerically computable form. However, the central difference formula

$$f' = \lim_{\hat{h} \to 0} \frac{f(\hat{x} + \hat{h}) - f(\hat{x} - \hat{h})}{2 \cdot |\hat{h}|} \tag{2}$$

produces results with up to two orders of magnitude better precision [6]. In order to achieve this high precision, it is important to choose the step size h so that it minimizes the round-off as well as the fractional error of the derivation computation. To this end,

$$\hat{h} = \sqrt[4]{eps} \cdot \hat{x} \tag{3}$$

has proven to be a suitable estimate for the central difference formula [6]. Here, $\hat{x}$ denotes the transition probability for the current stroke, while *eps* describes the smallest representable floating-point number. It is important to note that in cases, where $\hat{x}$ contains elements at the minimum/maximum of the allowed range of probabilities (0 / 1), it is impossible to apply the central difference formula. Hence the derivative is computed using the forward difference for the minimum values, and the backward difference

$$f' = \lim_{\hat{h} \to 0} \frac{f(\hat{x}) - f(\hat{x} - \hat{h})}{|\hat{h}|} \tag{4}$$

for the maximum values respectively. Like $\hat{x}$ the step size $\hat{h}$ is also a vector of length 36 describing the dimensions along which the derivative will be computed in the model-parameter-space. It represents the change in tactical behavior which is to be analyzed. Depending on the behavior which should be simulated, $\hat{h}$ also contains varying numbers of non-zero transition probabilities, which we will describe in the following sections.

### Partial Derivation

In mathematics, the partial derivative of a multi-variable function is its derivative with respect to a single of those variables, while the rest is considered constant. Applied to the Markov simulation this means that only a single transition probability will be derived, or in other words, $\hat{h}$ only contains one non-zero component for the respective transition. In this study we chose to examine the impact of point and error rates per stroke on the winning probability as one application of the partial derivation in table tennis. This is done by deriving the transitions to the absorbing states "point" and "error" for each stroke. The results of this simulation are shown in section 3.1.

As we already explained in section 1, however, all modifications to a transition probability have to be compensated in some way by the other transitions in that matrix row. Since we only derive one of the three available transitions per stroke in the partial derivation, there remain two other variables for compensation. In order to reach a uniform distribution of the tactical change over the remaining variables, we chose to compensate it using the formula given by Lames [3]:

$$\delta P_{yi} = -\frac{P_{yi}}{1 - P_x} \cdot \delta P_x \quad \text{for} \quad i = 1, \cdots, n \tag{5}$$

Here $P_{yi}$ describes the $n$ transition probabilities used to compensate for the change $\delta P_x$ in the investigated probability $P_x$. The compensation also means that although in theory the single change in error / point probability represents a partial derivative, the conclusive computation results in a directional derivative overall ($\hat{h}$ contains more than one non-zero element).

### Directional Derivation

The second derivation process examined in this study is the directional derivation in the direction of safer/riskier play. The tactical behavior "safer play" is thereby defined as a decrease in the transition probability to the states point of player A and point of player B at the current stroke. This roughly translates the fact that a player is less likely to commit errors if he plays with more caution, but at the same time also scores less direct points. The term "riskier play" is defined accordingly by an increase of the error and point transition probabilities, which in turn expresses that the error rate is expected to grow at the same rate as the direct point rate if a player plays more aggressively. To preserve the stochastic nature of the transition matrices, it is again necessary to compensate the change in transition probability that is induced by the derivation. To account for this, the transition probability to the next state (e.g. from the third stroke of player A to the fourth stroke of player B) will be de- or increased by the total amount of change, which is $|2\hat{h}|$ for the central difference formula and $|\hat{h}|$ for the forward and backward differences. The results of this simulation are shown in section 3.2.

**Winning Probability**

After adjusting the transition matrices, the new winning probability for each player can be calculated using the formula

$$f(x_1, \cdots, x_{36}) = P_{Win} = 0.5 \cdot (P_{P_{ServeA}} + P_{P_{ServeB}}) \tag{6}$$

for both methods. $P_{Win}$ denotes the point probability of a player in a rally with either player A or B as the serving player respectively. The winning probability is then computed from the point probabilities $((P_{P_{ServeA}}$ and $P_{P_{ServeB}})$ under the assumption that the total number of services is distributed evenly among both players. Although this does not entirely match the distribution of services between players, it gives a reasonable accurate estimation of service behavior over all games.

Algorithm 2.1 shows a summary of the simulation process in pseudo code, where $M$ represents a transition matrix as simulation input:

---

**Algorithm 2.1:** SIMULATION($M$)

---

**for each** *stroke* $\in M$

**do** $\begin{cases} \text{compute step size } \hat{h} \\ \text{adapt transition probabilities by } \hat{h} \\ \text{calculate new winning probabilities } f(\hat{x} - \hat{h}) \text{ and } f(\hat{x} + \hat{h}) \\ \text{compute central difference (equation 2)} \end{cases}$

---

## 3  Results

### 3.1  Partial derivation: Error and Winner rates

The partial derivation of point and error rates was used to study the impact of those probabilities on the overall winning probability as described in section 2.3. The resulting derivatives were then analyzed by a three-factor variance analysis with repeated measurements with the derivation direction (point/error) and the stroke number as within-subject effects and the gender as between-subject effect.

Tables 1 and 2 show the results of the conducted analyses. Since the data does not comply with the sphericity criteria, the Greenhouse-Geisser corrected values are considered for the interpretation of results. As the values in table 1 show,

Table 1. 3-factor variance analysis, within-subject tests (Greenhouse-Geisser)

| Source | Square sum (type III) | df. | Mean squares | F | Sig. | partial eta-square |
|---|---|---|---|---|---|---|
| Point | .040 | 1.000 | .040 | 1.275 | .259 | .002 |
| Point * Gender | .000 | 1.000 | .000 | .006 | .936 | .000 |
| Stroke Nr | 54.608 | 1.051 | 51.969 | 406,478 | .000 | .441 |
| Stroke Nr * Gender | 9.355 | 1.051 | 8.903 | 69.638 | .000 | .119 |
| Point * Stroke Nr | .300 | 1.699 | .176 | 15.608 | .000 | .029 |
| Point * Stroke Nr * Gender | .028 | 1.699 | .017 | 1.467 | .230 | .003 |

both the stroke number and the interaction of stroke number with gender and point/error probability show significant results. Thereby the stroke number holds the biggest influence on the variance, with a partial eta-square value of .441, followed by the combinations of stroke number with gender and point. Table 2 shows that the variance can also be explained by the gender with a significance of $p < .001$ and a partial eta square of .119. Lastly, figure 1 compares the

Table 2. 3-factor variance analysis, between-subject tests

| Source | Square sum (type III) | df. | Mean squares | F | Sig. | partial eta-square |
|---|---|---|---|---|---|---|
| Constant term | 399.568 | 1 | 399.568 | 11782.146 | .000 | .958 |
| Gender | 2.358 | 1 | 2.358 | 69.544 | .000 | .119 |
| Error | 17.499 | 516 | .034 | | | |

absolute mean values of the derivation of point and error probability between stroke number and gender. There exists a clear pattern in that the long rallies with more than five strokes generate derivatives with a higher absolute value than the rest of the strokes. With the absolute amount of change as important indicator for tactical behavior, the long rallies can be considered as the most impacting rallies in a game. This is especially true for the women's games, since the mean derivative over the long rallies is significantly higher than the same value of the men's data.

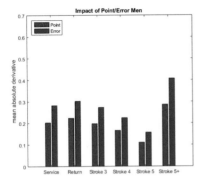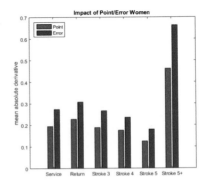

**Fig. 1.** Impact of Point and Error probability per stroke and gender

## 3.2   Directional derivation: Risk-taking

To examine the importance of the tactical behaviors "risky play" and "safe play", we conducted directional derivatives as described in section 2.3. The resulting derivatives were then visualized by the method described in the next section to identify potential tactical patterns for further analyses.

### Visualization

Since the visualization of multi-dimensional data is a difficult task, the number of displayed variables was reduced to three. To further examine the influence of risk and safety on the winning probability, we plotted the derivatives of each game over the point and error probability for each stroke as seen in figure 2. The

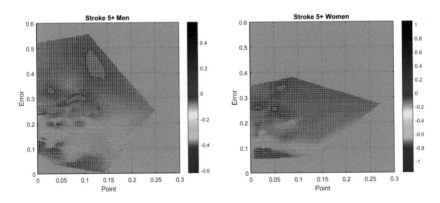

**Fig. 2.** Impact of Risk in long rallies (5+)

x- and y-axis represent the point and error probabilities respectively, while the color encodes the value of the derivative. To generate the shown topology, the derivation values of each game are interpolated on a fine-grained grid by Voronoi-interpolation between data points. Finally, an intersecting plane is drawn at the virtual "height" zero to highlight the difference between positive ande negative derivation values. On basis of the visualizations, the hypothesis was established that it is more beneficial for the losing player to take risks than for the winning player, since losing players showed a substantially higher amount of positive derivation values. Table 3 shows the results of the paired t-test for dependent samples for each stroke. The data shows a significant difference between the

**Table 3.** T-Test between Winner and Loser for all strokes

| Stroke | Mean | Std. Dev. | T | df | Sig.(2-sided) | r | Cohen's d |
|--------|------|-----------|---|----|--------------| --|-----------|
| Total | -0.154 | 0.201 | -30.25 | 1553 | .000 | -0.704 | -1.417 |
| 1 | -0.159 | 0.113 | -22.531 | 258 | .000 | -0.164 | -2.136 |
| 2 | -0.154 | 0.115 | -21.566 | 258 | .000 | -0.216 | -2.09 |
| 3 | -0.137 | 0.117 | -18.747 | 258 | .000 | -0.339 | -1.906 |
| 4 | -0.111 | 0.121 | -14.927 | 258 | .000 | -0.571 | -1.645 |
| 5 | -0.085 | 0.102 | -13.482 | 258 | .000 | -0.864 | -1.617 |
| 5+ | -0.276 | 0.395 | -11.265 | 258 | .000 | -0.940 | -1.379 |

means of the derivatives of the winning and losing player for each stroke. To compute the effect sizes for the dependent tests, the formula given by Dunlap et al. [1] was used. All effect sizes indicate a large effect of the difference between winner and loser, although the effect strength diminishes with growing stroke number. Furthermore the sign of the correlation $r$ validates our assumption, that losing players benefit from risky play, while it is overall detrimental for the winning players. As shown in section 3.1, the tactical behavior in the long rallies has an exceptional impact on the outcome of a game. Since we only examined the impact of single behaviors in the section before, we also performed a multiple regression analysis to expose any relation between the change in winning probability and risky/safe play for rallies lasting longer than five strokes (tables 4 and 5). For the regression we chose the point and error probabilities as predictors, and the derivative from the simulation as the dependent variable. The regres-

**Table 4.** Regression summary

| R | $R^2$ | Adjusted $R^2$ | Std. Error of the Estimate |
|---|-------|----------------|----------------------------|
| .444 | .197 | .194 | .219 |

Predictors: (Constant), E5+, P5+

Dependent Variable: D5+

sion summary reveals a rather small value with only 19.4% of the variance being defined by the predictors. However, the generated model can be trusted with values of $F = 63,230$ and $p < .001$ in an analysis of variance. Table 5 reveals the

**Table 5.** Regression - Coefficients

| Model | Unstd. Coefficients | | Std. Coefficients | | |
|---|---|---|---|---|---|
| | B | Std. Error | Beta | T | Sig. |
| (Constant) | -.278 | .027 | | -10.268 | .000 |
| P5+ | .197 | -.758 | -.121 | -3.069 | .002 |
| E5+ | 1.021 | .095 | .425 | 10.756 | .000 |
| Dependent Variable: D5+ | | | | | |

impact of each predictor on the change of winning probability. Both point and error probability have a significant effect on the winning probability. In comparison, the error probability ($Beta = .425$) contributes almost four times as much to the explanation of the variance as the point probability ($Beta = -.121$). This can also be identified in figure 2, where an almost linear increase in the derivative can be observed along the y-axis, which represents the error probability.

## 4    Conclusion

In conclusion, the method presented in this study is a conceptual improvement of the previous method by eliminating the need to explicitly model the difficulty of tactical behaviors in the context of Markov simulations. This allows the direct application of our method to different sports without knowledge of the difficulty of the involved tactics. Due to its numerical nature, the simulation is susceptible to errors like rounding and machine imprecision. With a careful selection of step size these difficulties can be overcome, however. We also showed that our method generates meaningful results when applied to tactical problems in table tennis. Future work on the method will implement an improved way to compute the winning probability by including the actual distribution of services between both players. The differences between winning and losing players also encourage further studies. Finally, a detailed comparison of the traditional and our method is planned to set both models into relation.

## References

1. Dunlap, W., Cortina, J., Vaslow, J., Burke, M.: Meta-analysis of experiments with matched groups or repeated measures designs. Psychological Methods 1 (2), p. 170–177 (1996)
2. Hohmann, A., Zhang, H., Koth, A.: Performance diagnosis through mathematical simulation in table tennis in left and right handed shakehand and penholder players. Science and Racket Sports III, p. 220–226 (2004)

3. Lames, M.: Leistungsdiagnostik durch Computersimulation [Performance diagnosis by computer simulation]. Harri Deutsch (1991)
4. Pfeiffer, M., Hohmann, A., Siegel, A., Böhnlein, S.: A markov chain model of elite water polo competition. Biomechanics and Medicine in Swimming XI, p. 278–280 (2010)
5. Pfeiffer, M., Hohmann, A., Zhang, H.: A mathematical-modeltheoretical concept to evaluate the performance relevance of tactical behaviour in the competition of young handball players. In: 8th Annual Congress European College of Sport Science (2003)
6. Press, W., Teukolsky, S., Vetterling, W., Flannery, B.: Numerical Recipes - The Art of Scientific Computing. Cambridge University Press (2007)

# Prediction and control of the individual Heart Rate response in Exergames

Katrin Hoffmann[1], Josef Wiemeyer[1], Sandro Hardy[2]

[1] Institute of Sport Science, Technische Universität Darmstadt, Germany
{hoffmann, wiemeyer}@sport.tu-darmstadt.de
[2] Multimedia Communications Lab (KOM) Technische Universität Darmstadt, Germany
sandro.hardy@kom.tu-darmstadt.de

**Abstract.** Setting an appropriate training load is one of the key elements for the success of exergames. Especially for the cardiovascular training, the adaptation of the current training load in accordance to an individually predetermined target training load plays an important role. In this paper, a new approach for the estimation and prediction of an individual's heart rate based on a monoexponential formula is presented and evaluated using statistical data. The estimation and prediction of the heart rate is a key factor for the calculation of adequate exertion parameters and therefore for the adaptation and personalization of exertion games, i.e. games that use whole-body exercises for game control. The tests reveal that the course of the heart rate response to changes of load bouts is not stable. Only a differential influence of gender on the HR course depending on the particular load bout can be found.

## 1    Introduction

In the physical training process, the appropriate strain of the human body is important to elicit optimal adaptations. However, the same defined external load can lead to different strain in different organisms. This individual response of the organism depends on a variety of factors, e.g., age, gender, BMI (i.e., ratio of body weight, and squared body height), fitness level, training biography and training goals. The response can be measured in various systems, e.g., the cardiovascular system (heart rate, blood pressure), the respiratory system (oxygen uptake), the metabolic system (lactate, ammonia), the hormonal system (cortisol, IGF-I), the immune system (leucocytes), or the autonomous system (adrenaline). An easy to measure indicator of the individual cardiovascular strain is the response of heart rate (HR in beats per minute, bpm) to the change of load bouts. In the sub maximal range, two characteristics of HR are important for estimating individual strain:

(a) The immediate short-term dynamics of HR after the onset of exercise or change of training load.

(b) The relation of HR and different training loads when HR has reached a steady state.

Bunc et al. [1] described the course of this response by an exponential equation:

© Springer International Publishing Switzerland 2016
P. Chung et al. (eds.), *Proceedings of the 10th International Symposium on Computer Science in Sports (ISCSS)*, Advances in Intelligent Systems and Computing 392, DOI 10.1007/978-3-319-24560-7_22

$$HR_{current} = a - b \cdot e^{-c \cdot t}. \tag{1}$$

Legend:

a    steady state HR level elicited by the change of load ($HR_{end}$ in Figure 1)

b    HR reserve, i.e., difference between a and the HR at the start of exercise

c    slope of HR curve

t    time [min]

Figure 1 illustrates the formula by means of a prototypical HR response.

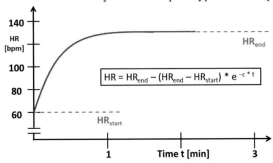

**Fig. 1.** Illustration of the time course of HR according to the Bunc equation (Bunc et al., 1988).

The issue of adequate strain is also relevant for the application of Exergames to aerobic training. These games measure a player's bodily movement with sensor technology and use this information to control a video game. Such games have the potential to be used as efficient and effective tool to enhance physical fitness, at least at low fitness levels [10]. So, it is important to provide proper training stimuli for those games to ensure training efficiency. In Exergames, the additional difficulty arises that control of training strain may interfere with game control. Taking the example of a competitive running game, the player may increase pace to overtake an opponent which may result in a HR that is too high for optimal adaptations.

The purpose of the study was to test the feasibility of the Bunc equation within an Exergame. In particular, the use of the equation for predicting and controlling current strain is evaluated. On that account, different parameters of the formula are tested with HR data that was obtained by measuring the HR response while participants were playing the Exergame "Letterbird" [3] [4]. This Exergame is based on the Exergame technology platform "StoryTecRT" [5]. Using a bike ergometer as input device, "LetterBird" aims at providing playful endurance training.

This paper addresses the following research questions:

1. Is the individual course of the HR response stable over different load conditions inside the Exergame "LetterBird"?
2. Can this individual response be described and predicted by selected individual influencing factors?

# 2    Methods

## 2.1    Participants

HR data of 16 healthy and active participants (9 men, 7 female, age: M = 26.6 yr., SD = 4.45, Range = 20 – 51 yr; BMI: M = 22.9, SD = 1.78, Range = 19.8 – 26.9) at the change of load bouts was used for the calculations. All participants reported to practice sport for at least 1.5 hours per week and to be active in their leisure time. 15 participants worked out more than 3 hours per week. 13 of 16 participants are members in a sports club. Demographical and anthropometric data is illustrated in Table 1.

**Table 1.**    Demographical and anthropometric description of the participants.

|  | Males (n=9) | | | Females (n=7) | | | Total (n=16) | | |
|---|---|---|---|---|---|---|---|---|---|
|  | M | SD | Range | M | SD | Range | M | SD | Range |
| **Age [yrs]** | 26.56 | 4.45 | 13.00 | 33.43 | 10.49 | 31.00 | 29.56 | 8.18 | 31.00 |
| **Height [m]** | 1.84 | 0.06 | 0.21 | 1.73 | 0.06 | 0.15 | 1.80 | 0.09 | 0.28 |
| **Weight [kg]** | 81.11 | 7.10 | 20.00 | 65.00 | 5.72 | 18.00 | 74.06 | 10.40 | 38.00 |
| **BMI [kg/m2]** | 23.70 | 1.67 | 5.20 | 21.81 | 1.37 | 4.20 | 22.88 | 1.78 | 7.10 |

## 2.2    Apparatus

All tests were performed on a cycle ergometer with a flywheel (Daum Ergometer TRS 3, Germany). The HR was monitored by a chest belt (Polar T31, Finland) and processed by the ergometer. This data was saved together with a corresponding time stamp, the load (P in Watt, W) and the pedal rate (in revolutions per minute, rpm). The ergometer was connected to a computer where the exergame "Letterbird" was running.

The goal of the exergame "LetterBird" is to collect letters that are approaching a pigeon in different altitudes. By varying cadence, the player controls the flight altitude of the pigeon: If the player pedals faster the pigeon rises, whereas the pigeon sinks down at lower pedal rates.

In this test, the game control was set to control the pigeon within a range of 60 rpm to 80 rpm. The pigeon was flying at the bottom of the screen at 60 rpm and on top of the screen at 80 rpm. An increase or decrease beyond this range did not influence the gameplay. According to Löllgen et al. [7], this variance of pedal rate was expected to have no significant influence on HR.

The individual HR response of the participant was measured in a study for evoking an individually optimal training HR ($HR_{intended}$). Therefore, the HR was obtained in two phases: the calibration and the exercise phase (see Figure 2) [6].

In the calibration phase, the individual HR response to two defined, successive load levels (two minutes each) was measured while the participants were playing the Exergame "LetterBird". The load was set depending on the body weight (in kg) of the participant:

$$\text{load level 1 } (L_1) \text{ with } L_1 = 1 \text{ W/kg} * \text{body weight} \qquad (2)$$

$$\text{load level 2 } (L_2) \text{ with } L_2 = 2 \text{ W/kg} * \text{body weight} \tag{3}$$

The obtained HR data was used to calculate a target load ($L_{target}$) that is expected to elicit $HR_{intended}$.

A short break after the calibration phase allowed the HR to return to the resting level before starting the exercise phase. The first four minutes of the exercise phase replicated the calibration phase to validate the data obtained. This time schedule was determined as standard procedure. If $L_2$ in calibration phase was considerably higher than $L_{target}$ the strain on the participant likely exceeded the sub maximal range. Therefore, $L_2$ in the exercise phase was reduced to $L_{target}$ ensuring a HR response resulting in a steady state. This time schedule was determined as exception procedure. Prototypical HR courses for each procedure are displayed in Figure 2.

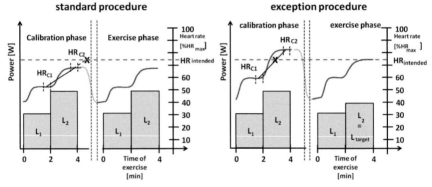

**Fig. 2.** Prototypical HR courses during the first 4 minutes of the calibration and exercise phase of the study. Left side: standard procedure, right side: exception procedure; solid line: measured HR, dashed line: $HR_{intended}$; $HR_{C1}/HR_{C2}$: last 30 sec of each load level used for calculation [6].

## 2.3   Procedure

**Examination of HR data.**
In this study, the first four minutes of each procedure were examined resulting in four load conditions:

- $L_1$ in the calibration phase:            $L_{1C}$
- $L_2$ in the calibration phase:            $L_{2C}$
- $L_1$ in the exercise phase:               $L_{1E}$
- $L_2$ (standard procedure) or $L_{target}$ (exception procedure) in the exercise phase:            $L_{2E}$

For every load bout, the measured HR was approximated using the Bunc equation. On that account, the data was first normalized. Parameters a and b were directly estimated from the individual HR data with a represented by the steady state HR elicited by the change of load and b represented by the difference between mean HR at the end of the load level and initial HR. Parameter c was estimated using the linear regression analysis.

Two load conditions were excluded from further calculation due to a negative c value. In one case, this negative value was caused by a measuring error and in the other case by a load reduction. This leads to a total number of 62 load conditions:

- 16 conditions in $L_{1C}$
- 16 conditions in $L_{2C}$
- 15 conditions in $L_{1E}$
- 15 conditions in $L_{2E}$

Six participants performed the standard procedure, whereas 10 participants performed the exception procedure.

**Stability test.**
The stability of the parameter c over the different load conditions was tested by calculating Pearson correlation coefficients:
1. Stability between phases (within load levels):
     (a) correlation of c_$L_{1C}$ and c_$L_{1E}$
     (b) correlation of c_$L_{2C}$ and c_$L_{2E}$
2. Stability between load conditions ($L_1$ and $L_2$):
     (a) correlation of c_$L_{1C}$ and c_$L_{2C}$ (calibration phase)
     (b) correlation of c_$L_{1E}$ and c_$L_{2E}$ (exercise phase)

**Factors influencing c.**
Factors influencing c were analyzed using different statistical methods. Age (in years), body weight (in kg) and physical activity (in hours per week) were correlated to the c values of all four load levels. Additionally, influences of gender were examined using a 2 (gender/activity level) x 2 (phase) x 2 (load level) ANOVA with repeated measures.

# 3 Results

## 3.1 Stability test

The stability test of the parameter c revealed unsatisfactory reliability (see Table 3). At equal load levels the correlations range from -0.08 to 0.34. No correlation is significant.

**Table 2.** Retest reliability of c for different load levels ($L_1$ and $L_2$), phases (C – calibration; E – exercise), and procedures

| Correlation | Total | Standard procedure | Exception procedure |
|---|---|---|---|
| 1. (a) c_$L_{1C}$ - c_$L_{1E}$ | 0.34 | | |
| 1. (b) c_$L_{2C}$ - c_$L_{2E}$ | 0.12 | 0.30 | -0.08 |
| 2. (a) c_$L_{1C}$ - c_$L_{2C}$ | -0.36 | | |
| 2. (b) c_$L_{1E}$ - c_$L_{2E}$ | -0.41 | -0.59 | -0.44 |

## 3.2    Factors influencing c

The influence of body weight on c is inconsistent and small (see Table 3). Only one correlation is significant. Body weight is correlated significantly positive with $c\_L_{2C}$. This means that the higher the body weight the steeper the HR rises at load level 2.

The influence of age is also inconsistent and small (see Table 4). None of the correlations are significant.

No significant correlations were found between activity level and c (see Table 5).

**Table 3.** Correlation between body weight (BW) and c for different load levels ($L_1$ and $L_2$), phases (C – calibration; E – exercise), and procedures

| Correlation | Total | Standard procedure | Exception procedure |
|---|---|---|---|
| **BW – c_$L_{1C}$** | -0.380 | | |
| **BW – c_$L_{1E}$** | -0.360 | | |
| **BW – c_$L_{2C}$** | 0.548 * | | |
| **BW – c_$L_{2E}$** | 0.042 | 0.232 | -0.060 |

\* $2p < 0.05$

**Table 4.** Correlation between age and c for different load levels ($L_1$ and $L_2$), phases (C – calibration; E – exercise), and procedures

| Correlation | Total | Standard procedure | Exception procedure |
|---|---|---|---|
| **Age – c_$L_{1C}$** | 0.396 | | |
| **Age – c_$L_{1E}$** | 0.202 | | |
| **Age – c_$L_{2C}$** | -0.179 | | |
| **Age – c_$L_{2E}$** | -0.277 | -0.382 | -0.291 |

**Table 5.** Correlation between activity level (AL; hours per week) and c for different load levels ($L_1$ and $L_2$), phases (C – calibration; E – exercise), and procedures

| Correlation | Total | Standard procedure | Exception procedure |
|---|---|---|---|
| **AL – c_$L_{1C}$** | 0.081 | | |
| **AL – c_$L_{1E}$** | -0.241 | | |
| **AL – c_$L_{2C}$** | 0.268 | | |
| **AL – c_$L_{2E}$** | -0.083 | -0.586 | 0.307 |

Gender differences are illustrated in Figure 3. 2x2x2 ANOVA with repeated measures revealed a significant two way interaction of gender and load level ($F1, 12 = 5.17$, $p<.05$, η2part $=0.301$). No main effects or further interactions were significant.

According to Figure 3, males had significantly lower c values at load level 1 both in the calibration phase (U test: $z = -1.75$, $p<0.05$) and in the exercise phase (U test: $z = -2.03$, $p<0.05$), whereas females had significantly lower c values at load level 2 in the calibration phase (U test: $z = -2.23$, $p<0.05$).

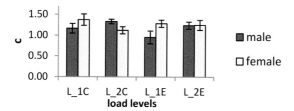

**Fig. 3.** Comparison of mean c value for men and women in the corresponding load levels.

## 4    Discussion

The non-significant correlations in the stability tests revealed that the c values are highly dependent on the particular load level and phase. This leads to the conclusion that the isolated c value is not useful to predict HR courses in the Exergame "Letterbird". Further differentiation of the participants is needed.

On that account, the influences of age, gender, body weight, and activity level were tested for all load changes. The data revealed that there are no consistent correlations between age, body weight, and activity level and the c values. There was only one significant correlation between body weight and $c\_L_{2C}$. All other correlations were not significant. Concerning body weight and age, correlations even changed sign.

Gender was a factor discriminating among c values. However, the significant interaction of gender and load level revealed a slower HR increase in males at the lower load level (irrespective of phase), whereas HR increase was steeper in males at the higher load level in the calibration phase.

The following statements for a prediction of the HR response to defined load changes can be made using the present data:

- The increase of HR at the onset of exercise and change of load is not stable over different load conditions inside the Exergame "LetterBird".
- Age, body weight, and activity level cannot be used as indicators for predicting the HR course.
- Gender differences have to be considered depending on load level.

A possible reason for the lack of stability of c is the influence of pedal rate on HR. Due to motivating factors, the participants exceeded the pedal rate beyond the given range to control the game. This became especially apparent when the controlled pigeon was located far apart from the collectable letter. A more promising factor for individual load control may be the performance level of the participants.

Furthermore, the sample size in this study is very small and effects need to be re-tested using a bigger sample size. Additionally, the effect of age needs to be tested in a wider range, as the sample is not covering the whole range of age (range: 20 – 51years). Particularly in adults, the increase of HR should be steeper compared to children (e.g., [2], [9]).

In future studies, more sophisticated approximation algorithms (i.e. PerPot [8]) for predicting the HR course need to be considered.

## 5    Conclusion

A prediction of HR within the Exergame "LetterBird" using solely the Bunc equation
is considered difficult as the slope of HR was not stable at different load levels and
phases. Further tests revealed only a differential influence of gender depending on the
particular load level.

## 6    Acknowledgements

We thank the Forum Interdisziplinaere Forschung at Technische Universität Darmstadt for financial funding. Additionally, we thank the Multimedia Communication
Laboratory (KOM) at Technische Universität Darmstadt for the resources provided.

## References

1. Bunc, V.P., Heller, J. & Leso, J.: Kinetics of heart rate response to exercise. Journal of
   Sports Science, 6 (1):39-48 (1988).
2. Cooper, D. M., Berry, C., Lamarra, N. & Wasserman, K.: Kinetics of oxygen uptake and
   heart rate at onset of exercise in children. Journal of Applied Physiology 59 (12), 211-217.
   (1985).
3. Hardy, S., Dutz, T., Wiemeyer, J., Göbel, S., & Steinmetz, R. (2014). Framework for personalized and adaptive game-based training programs in health sport. Multimedia Tools
   and Applications, 1-23.
4. Hardy, S., Göbel, S., Gutjahr, M., Wiemeyer, J. & Steinmetz, R.: Adaptation Model for
   Indoor Exergames. International Journal of Computer Science in Sport.11 (SE). (2011).
5. Hardy S., Göbel, S. & Steinmetz, R.: Adaptable and Personalized Game-based Training
   System for Fall Prevention. Technical Demo Paper. In Proceedings of the 21st ACM International Conference on Multimedia (Barcelona, Catalunya, Spain, October 21 – 25, 2013)
   ACM New York, NY. (2013).
6. Hoffmann, K., Wiemeyer, J., Hardy, S. & Göbel, S.: Personalized adaptive control of
   training load in Exergames from a sport-scientific perspective: Towards an algorithm for
   individualized training. In S. Göbel & J. Wiemeyer (eds.), Games for Training, Education,
   Health, and Sports, 129-140 (2014).
7. Loellgen H, Graham, T, Sjogaard, G.: Muscle metabolites, force, and perceived exertion
   bicycling at varying pedal rates. Medicine and Science in Sports and Exercise; 12(5), 345–
   351 (1980).
8. Perl, J. (2001). PerPot: A metamodel for simulation of load performance interaction. European Journal of Sport Science, 1(2), 1-13.
9. Springer, C., Barstow, T.J., Wasserman, K. & Cooper, D.M.: Oxygen uptake and heart rate
   responses during hypoxic exercise in children and adults. Medicine and Science in Sport
   and Exercise, 23 (1), 71-79 (1991).
10. Wiemeyer, J. & Kliem, A.: Serious games and ageing - a new panacea? European Review
    of Aging and Physical Activity, 9 (1), 41-50 (2012).

# Evaluation of changes in space control due to passing behavior in elite soccer using Voronoi-cells

Robert Rein[1], Dominik Raabe[1], Jürgen Perl[2], Daniel Memmert[1]

[1]Institute of Cognition and Team/Racket Sport Research, German Sport University Cologne
[2]Institute of Computer Science, FB 08, University of Mainz, Germany

## 1    Introduction

A soccer player's ability to make an "effective" pass in a play situation is considered one of the key skills characterizing successful performance in elite soccer (Bush, Barnes, Archer, Hogg, & Bradley, 2015; Hughes & Franks, 2005; Mackenzie & Cushion, 2013). However, although there is ample evidence in the literature that passing behavior is important it is much less clear what actually characterizes a "good" pass. One common topic investigated with respect to passing behavior in elite soccer is the frequency of passing events and their correlation with game performance (Lago-Peñas, Lago-Ballesteros, Dellal, & Gómez, 2010; Liu, Gomez, Lago-Penas, & Sampaio, 2015). Results from this line of research show for example that the number of passes made during the FIFA World cup finals between 1966 and 2010 continuously increased (Wallace & Norton, 2014) and that the number of crosses and assists are positively related to game performance (Lago-Peñas et al., 2010; Luhtanen, Belinskij, Häyrinen, & Vänttinen, 2001). Similar, Hughes and Franks (2005) found that longer passing sequences increase the goal scoring likelihood (see also Yiannakos & Armatas, 2006). Although all these studies provide valuable information regarding the importance of passing behavior in elite soccer it remains unclear what characterizes successful passing. Accordingly, it is not known which effects individual passes exert on concrete game play situations which impedes the application of research findings by practitioners. Here, we introduce a novel approach to evaluate individual passing behavior during game play using Voronoi cells to investigate changes in space control.

The assumption behind this novel approach is that an effective pass gives the attacking side an advantage with respect to the control of space in front of the goal (Ensum, Pollard, & Taylor, 2004; Pollard, Ensum, & Taylor, 2004). Accordingly, a pass which increases space dominance of the attacking team is considered advantageous. To be able to evaluate changes in space control due to passing behavior however, an appropriate model of space dominance by the attacking team must be found. Previously, several methods to quantify the control over pitch space have been proposed (Fonseca, Milho, Travassos, & Araujo, 2012; Kim, 2004; Nakanishi, Murakami, &

© Springer International Publishing Switzerland 2016
P. Chung et al. (eds.), *Proceedings of the 10th International Symposium on Computer Science in Sports (ISCSS)*, Advances in Intelligent Systems and Computing 392, DOI 10.1007/978-3-319-24560-7_23

Naruse, 2008; Taki & Hasegawa, 2000). Conceptually, all these approaches are variations of the so-called Voronoi diagram.

A Voronoi diagram is a partitioning of a plane into different cells according to the distances between points in the plane. Each point, also called seeds, is associated with a single unique cell and the geometry of the cell is chosen such that all points within a given cell are closer in terms of the Euclidean distance to the seed associated with that cell than to any other seed. The individual cells are called Voronoi cells. Applied to soccer, the pitch constitutes the plane and each player represents a seed respectively. Each player can therefore be associated with a unique cell containing all pitch locations which are close to this player compared to all other players (compare Figure 1).

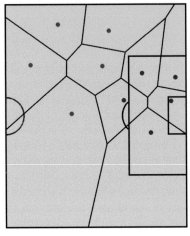

**Fig. 1.** Voronoi diagram of a player configuration: ▮ Team A, ▮ Team B

The player therefore controls the region contained in that cell. Naturally, this model simplifies the true situation as it assumes that all players are equally fast and neglects the current movement directions of the players. Early applications of this approach have been made by Taki and Hasegawa (2000) who used the Voronoi diagram. The authors did not use the standard Voronoi tessellation based on the Euclidean distance but defined a constant velocity function based on the concept of reachability and the finite running velocity. Result showed that the attacking team occupies a greater area compared to the defending team (compare also Gudmundsson & Wolle, 2014; Taki & Hasegawa, 2000). Fonseca et al. (2012) used Voronoi diagrams to investigate space control in experimental Futsal games. The results showed that the attacker team dominated more space compared to the defending team.

To obtain an assessment of passing efficiency, in the present work we combined therefore passing behavior with changes in space control using a Voronoi diagram. We expected that according to the region on the pitch from where the pass was made and to which region the pass was directed different effects on the space dominance of the attacking team would be visible.

## 2    Methods

In total 12 first halve-time games from the German first Bundesliga during the season 2012/2013 were analyzed. The dataset consisted of the x-y position data for each player and the ball recorded at 1Hz. The soccer pitch was divided into three areas: defensive third, mid-field, offensive-third (Tenga, Ronglan, & Bahr, 2010) to allow categorization of pass origin and pass target. Passing efficiency was calculated according to the change in space dominance of the attacking team in the 30m attacking zone. To this end the Voronoi diagram at pass initiation and at pass completion were calculated. Subsequently, the space dominance (in %) of the attacking and defending teams at these two time-point were determined. Finally, the attacking team space dominance at pass completion was subtracted from space dominance at pass initiation. All statistical tests were performed using the R statistical software. To take into account varying number of games analyzed for the teams a mixed-effects model was used with game and team as random-effects. In addition, non-parametric statistics were used where appropriate. The alpha level was set at $\alpha = 0.05$.

## 3    Results

On average relatively small changes were observed for changes in the percentage of the attacking areas dominated by the attacking team, $space_{30} = 1\% \pm 7\%$. However, as visible In Figure 2 this is mainly a result of the large fluctuations of the measure. Nevertheless, some general trends are immediately visible.

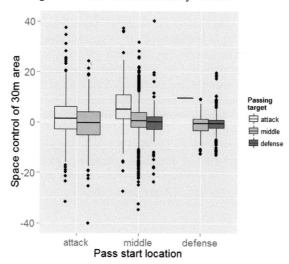

**Fig. 2.** Changes in space control of the 30m area as a function of pass start location and pass target location.

The plot indicates a trend that passes initiated from the mid-field lead to an increase in space dominance when targeting the attacking third whereas backwards passes into the defensive third decrease space dominance. Passes starting from the attacking area show a weak decrease in space dominance when targeting the mid-field and appear to increase space dominance when staying with the attacking area. In contrast, passes from the defensive third do not seem to influence attacking dominance. Statistical analysis indicated a significant main effect for passes made for from the mid-field, $\chi^2(2) = 156.7$, $p < 0.001$, with significant post-hoc tests for mid-field to attacking area passes, $z = 12.06$, $p < 0.001$ and within the middle area, $z = 3.7$, $p < 0.001$. Passes from the attacking area showed a significant main effect, $\chi^2(1) = 8.8$, $p < 0.01$, with significant effects for passes within the attacking area, $z = 3.5$, $p < 0.01$. No significant effects were found for passes from the made from the defensive third.

## 4    Discussion

Here we introduced a novel assessment method to evaluate the effectiveness of individual passes using changes in space dominance of the attacking area by the attacking team. The results show that passes made from the mid-field as well as from the attacking field on average provide the largest gain in space dominance for the attacking team although individual pass space gains vary widely. Thus, this indicates that the application of the present approach for team average measurements might be of limited value to assess passing effectiveness with respect to space dominance. However the approach provides however some interesting venues for future research. For example, rather than investigating the passing behavior of the whole team the present approach can be used to evaluate individual contributions to team passing behavior. A potential application could be to evaluate opponent's passing efficiency in preparation of team tactics for upcoming games. The approach allows to easily identify opponent's key players and their typical passing behavior in particular as the evaluation can be made automatically once passing events are identified in the data. The present method could also be combined with previous approaches (Gudmundsson & Wolle, 2014) to evaluate simulated passing opportunities versus actual made opportunities to improve decision making in players.

In summary, in the present work we introduced a novel approach to evaluate passing efficiency by analyzing the effect of passing behavior on changes in space domination of the attacking area. The approach taken might also be valuable to study other team sports like field and ice hockey or American football.

## References

Bush, M., Barnes, C., Archer, D. T., Hogg, B., & Bradley, P. S. (2015). Evolution of match performance parameters for various playing positions in the English Premier League. *Human Movement Science, 39*, 1-11. doi: 10.1016/j.humov.2014.10.003

Ensum, J., Pollard, R., & Taylor, S. (2004). Applications of logistic regression to shots at goal in association football: calculation of shot probabilities, quantification of factors and player/team. *Journal of Sports Sciences, 22*(6), [np].

Fonseca, S., Milho, J., Travassos, B., & Araujo, D. (2012). Spatial dynamics of team sports exposed by Voronoi diagrams. *Human Movement Science, 31*(6), 1652-1659. doi: 10.1016/j.humov.2012.04.006

Gudmundsson, Joachim, & Wolle, Thomas. (2014). Football analysis using spatio-temporal tools. *Computers, Environment and Urban Systems, 47*(0), 16-27. doi: http://dx.doi.org/10.1016/j.compenvurbsys.2013.09.004

Hughes, M., & Franks, I. (2005). Analysis of passing sequences, shots and goals in soccer. *Journal of Sports Sciences, 23*(5), 509-514. doi: 10.1080/02640410410001716779

Kim, S. (2004). Vornoi analysis of a soccer game. *Nonlinear Analysis: Modelling and Control, 9*(3), 233-240.

Lago-Peñas, Carlos, Lago-Ballesteros, Joaquín, Dellal, Alexandre, & Gómez, Maite. (2010). Game-Related Statistics that Discriminated Winning, Drawing and Losing Teams from the Spanish Soccer League. *Journal of Sports Science and Medicine 9*, 288-293.

Liu, H., Gomez, M. A., Lago-Penas, C., & Sampaio, J. (2015). Match statistics related to winning in the group stage of 2014 Brazil FIFA World Cup. *Journal of Sports Science, 33*(12), 1205-1213. doi: 10.1080/02640414.2015.1022578

Luhtanen, Pekka, Belinskij, Antti, Häyrinen, Mikko, & Vänttinen, Tomi. (2001). A comparative tournament analysis between the EURO 1996 and 2000 in soccer. *International Journal of Performance Analysis in Sport, 1*(1), 74-82.

Mackenzie, R., & Cushion, C. (2013). Performance analysis in football: a critical review and implications for future research. *Journal of Sports Science, 31*(6), 639-676. doi: 10.1080/02640414.2012.746720

Nakanishi, Ryota, Murakami, Kazuhito, & Naruse, Tadashi. (2008). Dynamic Positioning Method Based on Dominant Region Diagram to Realize Successful Cooperative Play. In U. Visser, F. Ribeiro, T. Ohashi & F. Dellaert (Eds.), *RoboCup 2007: Robot Soccer World Cup XI* (Vol. 5001, pp. 488-495): Springer Berlin Heidelberg.

Pollard, R., Ensum, J., & Taylor, Samuel. (2004). Estimating the probability of a shot resulting in a goal: The effects of distance, angle and space. *International Journal of Soccer and Science, 2*(1), 50-55.

Taki, T., & Hasegawa, J. (2000, 2000). *Visualization of dominant region in team games and its application to teamwork analysis.* Paper presented at the Computer Graphics International, 2000. Proceedings.

Tenga, A., Ronglan, Lars T., & Bahr, Roald. (2010). Measuring the effectiveness of offensive match-play in professional soccer. *European Journal of Sport Science, 10*(4), 269-277. doi: 10.1080/17461390903515170

Wallace, J. L., & Norton, K. I. (2014). Evolution of World Cup soccer final games 1966-2010: game structure, speed and play patterns. *J Sci Med Sport, 17*(2), 223-228. doi: 10.1016/j.jsams.2013.03.016

Yiannakos, A., & Armatas, V. (2006). Evaluation of the goal scoring patterns in European Championship in Portugal 2004. *International Journal of Performance Analysis in Sport, 6*(1), 178-188.

# What is the best fitting function? Evaluation of lactate curves with common methods from the literature

Stefan Endler[1], Christian Secker[1], and Jörg Bügner[2]

[1] Johannes Gutenberg University Mainz, Computer Science Department, 55122 Mainz, Germany
[2] German Olympic Sports Confederation (DOSB), 63263 Neu-Isenburg, Germany

**Abstract.** Using the lactate threshold for training prescription is the gold-standard, although there are several open questions. One open question is: What is the best fitting method for the load-lactate data points? This investigation re-analyses over 3500 lactate diagnostic datasets in swimming. Our evaluation software examines six different fitting methods with two different minimization criteria (RMSE and SE). Optimization of parameters of the functions is put in exccecution with gradient descent. From a mathematical point of view, the double phase model, which consists of two linear regression lines, shows the least errors ($RMSE_{min}$ 0.254 ± 0.172; $SE_{min}$ 0.311 ± 0.210). However, this method cannot be used for every further determination of lactate thresholds. Some threshold determination models need a single curve. In these cases, the exponential function shows the least errors ($RMSE_{min}$ 0.846 ± 0.488; $SE_{min}$ 1.196 ± 0.689). This confirms the default fitting method used in practice.

## 1   Introduction

Training processes in most sports depend on intensity, duration and volume of training [15]. Particularly, intensity is an individual variable for every athlete. A speed target for an athlete can be optimal whereas the same speed target could lead in exhaustion for another athlete, although both athletes have the same proficiency level. Therefore, individual optimal intensity for training prescription is determined with parameters such as $\%VO_2max$ or %heart rate reserve [15]. Most competitive athletes are using training intensities based on an individual threshold. One possibility is the determination of the anaerobic threshold using lactate measurement during an incremental step test [3]. If one do exercise above the anaerobic threshold, more lactate is build than can be eliminated at the same time. This would lead to an delayed exhaustion due to acidaemia. Therefore, the threshold is an important value for training prescription particularly in endurance sports. Despite many criticism about this determination method, today the lactate threshold (LT) is the gold-standard [6].

LT determination needs invasive measurement of blood lactate during an incremental step test. Afterwards, load values such as speed and lactate concentration

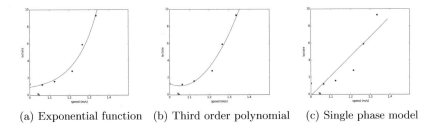

(a) Exponential function    (b) Third order polynomial    (c) Single phase model

**Fig. 1.** Approximation of speed-lactate points using an one curve

at every load level are plotted in a diagram. Based on these points in the diagram, a fitting curve is determined to get a continuous load-lactate relation. Based on that curve, LT can be determined by various models, such as OBLA [11] which defines LT at a fixed lactate concentration of 4 mmol/l. Using two different curves which differs a bit, LT determination by any model would result in different thresholds.

There is still a great debate on the best fitting procedure. Faude et al. [6] identified six different methods for load-lactate curve fitting in the literature. The fitting curve is essential important, because all LT determination algorithms are using the curve.

From a mathematical point of view, the best fitting curve has the least error between measured and determined load-lactate points. This investigation reanalyses a great dataset of lactate diagnostics in swimming. We examine the six identified methods from literature with objective error criteria.

### 1.1  Functions

The fitting functions so far published can be divided into two types. The first type tries to fit the data points by use of only one curve. The other type uses at least two curves for approximation.

The first type includes the exponential function [9] (1), the third order polynomial [5] (2) and the single phase model [6] (3). The functions are shown in figure 1, exemplary. The formulas of the three functions look as follows:

$$y = a + b \cdot x^{cx} \qquad (1)$$
$$y = a \cdot x^3 + b \cdot x^2 + c \cdot x + d \qquad (2)$$
$$y = a \cdot x + b \qquad (3)$$

where x and y are load values and lactate concentration, respectively. As parameter d in equation 2, the basis lactate concentration can be chosen, if it is measured before the incremental step test. The parameters a, b and c in all equations have to be estimated by minimizing an error criterion.

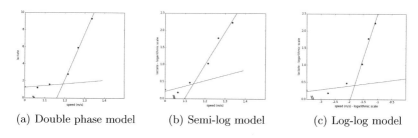

(a) Double phase model (b) Semi-log model (c) Log-log model

**Fig. 2.** Approximation of speed-lactate points using more than one curve

The exponential function was introduced by Hughson et al. [9]. They used a combination of the residual sum squares (RSS) and the correlation coefficient (r) as error criteria, whereat RSS has to be minimized and r has to be maximized. The exponential function produces a strictly monotonic increasing curve. In some cases the lactate concentration decreases at the first step(s) and increases only at later steps. This cannot be mapped by the exponential function and hence is a disadvantage of this fitting curve.

The third order polynomial introduced by Cheng et al. [5] was originally called curvilinear regression. Cheng et al. used a correlation coefficient as maximization criterion. In contrast to the exponential function, this function can also map decreasing lactate concentrations at the beginning, as one can see from figure 1(b).

The single phase model is mentioned in [6] and in combination with a semi-log and a log-log model in [2]. It should therefore be examined, even if the probability of success being the best model is very low. Particularly, because Cheng et al. [5] asserted, that first and second order regressions generate lower correlations compared to third order regressions.

The second category includes the double phase model [8], the semi-log model [2] and the log-log model [2, 10]. These methods are shown in Figure 2, exemplary.

The double phase model was introduced by Grant et. al. [8]. It uses two linear regression functions (3). The basic idea is a better fitting of the data points below and above the threshold. In the first phase a linear regression function is estimated only for the baseline lactate concentrations and in the second phase only for the incremental lactate concentrations. The data points which belong to the baseline regression are identified by a comparison of residuals. Normally, this method is used for a directly determination of LT. In this case, LT is defined at the intercept of both regression lines. Otherwise, it is absolutely possible to use the fitting for other LT determination methods such as OBLA [11].

Just as the double phase model, the semi-log model introduced by Beaver et al. [2] uses two linear regression functions. Unlike the double phase model, the data is transformed to an other scale, before estimating the regression parameters. In the semi-log model, the lactate concentrations are transformed by a natural logarithm. The decision which data point belongs to the baseline or incremental part, respectively cannot be made automatically and has to be made by visual inspection.

Compared to the semi-log model, in the log-log model both, lactate concentration and load values are transformed by natural logarithm.

## 2    Methods

### 2.1    Data

For our evaluation, we re-analyzed 3579 lactate tests (1881 male tests, 1698 female tests) of A-C team athletes in swimming. This collection includes all lactate diagnostics which were executed at the Olympic Training Centre in Hamburg[1] in the years from 1992 until 2006. Every athlete executed the test between two and four times a year at well-defined points in time. There is a high validity of this dataset, because physician, medical support personnel and training scientists were changeless over the whole period.

All athletes executed the lactate test by Pansold [12, 13]. It is a swimming-specific field test for diagnosing the endurance capacity, accounting for the different structures of swimming disciplines. The DSV is using the Pansold protocol since the beginning of the 1990s. The test is carried out for 400m in four steps and for 100m and 200m disciplines in five steps. The test is executed in a 50m pool, usually in the athletes main event. The load specification is determined by a percentage of the individual best, whereby the rest periods between the steps are fixed [1]. The step duration reduces from step to step, because of the constant distance and increasing swimming speed. Lactate concentration of the capillary blood is measured after every step with blood samples from the ear lobe.

### 2.2    Evaluation

We implemented a software to determine objective differences between the real measured lactate data and all fitting functions introduced above. The abstract algorithm is presented in table 1 as pseudo code. The algorithm is designed as a template method [7] inside an abstract class, so that it can be used for all fitting functions and with different error functions, easily.

The algorithm is implemented in Java 8 by use of the IDE Eclipse Kepler and can therefore run on every system. For every fitting function, an own subclass is implemented with specific implementations of parts of the algorithm.

The input function (*read()*) in the first step is implemented inside the abstract class, because this can be done independent of the fitting function. The data is stored in a set (*lactate_data*) for further determinations.

In the for-loop (steps 3-6) every series of lactate test data points is fitted by a

---

[1] We thank the Olympic Training Centre Hamburg/Schleswig-Holstein making the data available for us

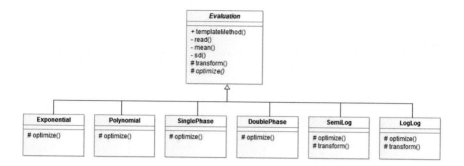

**Fig. 3.** Class diagram of the evaluation functionality

**Table 1.** Template method for the algorithm of determining mean and standard deviation of fitting function errors

| | |
|---|---|
| 1: | lactate_data ← read(Excel dataset) |
| 2: | Initialize min_error[#datasets] |
| 3: | **for** i=0 **to** #datasets **do** |
| 4: | lactate_data[i] ← transform(lactate_data) |
| 5: | min_error[i] ← optimize(lactate_data[i], errorFunction()) |
| 6: | **end for** |
| 7: | **output** mean(min_error), sd(min_error) |

curve and the minimum error is stored. Every lactate dataset itself has five or six data points (speed and lactate concentration) dependent on the swimming discipline of the athlete. The transformation function (*transform()*) in the fourth step is implemented as a hook. Subclasses have the possibility to override this method to transform the data to an other scale. This is needed for the semilog and the log-log model, i.e. it is only implemented in the subclasses *SemiLog* and *LogLog*. The optimization function (*optimize()*) in the fifth step optimizes the parameters of the fitting function. An optimal set of parameters can be determined by minimizing the distances (errors) between real measurements and fitting values. The distance is calculated by an error function (*errorFunction()*), which is assigned to the optimization function, too. The optimization function is implemented in special subclasses for every fitting method (see figure 3). The last step uses all minimal error values and calculates the mean (*mean()*) and the standard deviation (*sd()*).

We implemented two different error functions. First, the root mean square error (RMSE):

$$RMSE = \sqrt{\frac{1}{n} \cdot \sum_{i=1}^{n} (y\_real_i - y\_fitted_i)} \qquad (4)$$

**Table 2.** Mean $RMSE_{min}$ and $SE_{min}$ of different fitting models to lactate data

| Fitting Model | Mean $\pm$ SD of $RMSE_{min}$ (4) | Mean $\pm$ SD of $SE_{min}$ (5) |
|---|---|---|
| Exponential function | 0.846 $\pm$ 0.488 | 1.196 $\pm$ 0.689 |
| Third order polynomial | 1.388 $\pm$ 0.577 | 2.396 $\pm$ 0.997 |
| Single phase | 2.946 $\pm$ 0.887 | 3.322 $\pm$ 0.887 |
| Double phase | 0.254 $\pm$ 0.172 | 0.311 $\pm$ 0.210 |
| Semi-log | 0.254 $\pm$ 0.172 | 0.311 $\pm$ 0.210 |
| Log-log | 0.365 $\pm$ 0.247 | 0.448 $\pm$ 0.303 |

where n is the number of data points, y_real are the measured lactate concentrations and y_fitted are the approximated lactate concentrations by the fitting function. Secondly, the adapted root mean square error by Busso [4] (SE):

$$RMSE = \sqrt{\frac{1}{n - df - 1} \cdot \sum_{i=1}^{n}(y\_real_i - y\_fitted_i)} \qquad (5)$$

This error function considers the degree of freedom of the fitting function (df), that is the number of parameters, which have to be optimized. Thereby, the error values increase compared to RMSE dependent on the difficulty of optimizing the function parameters. This error function models the efficiency of generating the optimal curve besides the differences of measured and determined values.

As optimization algorithm, we decided to use gradient descent [16]. The optimization function returns the minimal error, which was produced by the optimal parameter set.

After all minimal errors are determined and collected in an array (line 2 and 5), the overall mean and standard deviation can be determined using the array of all minimal errors (line 7).

## 3   Results

The results of the evaluation with the software using all lactate diagnostics of the Olympic Training Centre are shown in table 1:

The least errors are produced by the double phase model and the semi-log model using $RMSE_{min}$ (0.254 $\pm$ 0.172) as well as $SE_{min}$ (0.311 $\pm$ 0.210). Both methods produce the same error, because both are using two linear regression functions. The only difference is the different scale of lactate values in the semi-log model. But, in this case, all values are transformed to normal scale before estimating the error value to ensure comparability.

The worst errors are produced by the single phase model ($RMSE_{min}$ 2.946 $\pm$ 0.887; $SE_{min}$ 3.322 $\pm$ 0.887).

The least errors considering solely methods which produce a single curve is the exponential function ($RMSE_{min}$ 0.846 $\pm$ 0.488; $SE_{min}$ 1.196 $\pm$ 0.689).

# 4  Discussion

An open question in lactate diagnostic is the optimal fitting method for load-lactate data points [6]. Based on the fitting curve, LT is determined by various models, e.g. OBLA [11]. Because of a great collection of lactate diagnostic data (>3500 datasets) in swimming from the Olympic Training Center in Hamburg, this investigation can give an objective answer to the open question.

There are two different types of fitting methods. One type fits the data points by single curves, others try to fit the points by two or more regression lines.

Considering only the three models which uses a single curve (exponential function, third order polynomial and single phase), the exponential functions produces the least errors. This function is used in practice by default. If one wants to approximate all data points by a single curve, our investigation confirms the exponential function as default. A single curve is needed for various LT determination models, such as the individual anaerobic threshold by Simon [14], which defines the threshold at the 45° tangent to the lactate curve. However, an exponential function cannot map the first steps of a lactate diagnostic, if lactate concentration decreases firstly. This is due to exponential functions are strictly monotonic increasing provided that parameters b and c in (1) are positive. In cases of firstly decreasing lactate concentrations, the polynomial function could produce better results, because this function is not strictly monotonic increasing (see figure 2). It would be interesting to evaluate both methods again using only lactate diagnostics, which show a firstly decreasing lactate concentration.

From a pure mathematical point of view, using more than one function result in lower differences between measured and approximated lactate concentrations. The idea is to separate the baseline and the increasing phase of the load-lactate behaviour [8]. Within the three models, the double phase model and the semilog model produce the least errors. The adaptation of the lactate concentrations to a logarithmic scale has no effect, so that this additional computation can be stinted. I.e. the optimal fitting method for load-lactate datasets is the double phase model. The disadvantage with this model is the further use of the fitting curve for LT determination. Using this method limits the LT determination models. Some models can be used with the double phase method, such as OBLA which defines LT at a fixed lactate concentration. Others, such as the Simon method [14] cannot be determined by use of the double phase model. This LT determination model needs a single curve.

Generally, it should be mentioned, that the lactate diagnostic procedure by Pansold [12, 13] produces special datasets. Lactate diagnostics in running or cycling uses fixed steps in the graded incremental test. This produces equidistant intervals in contrast to the procedure by Pansold. Therefore, all results are possibly limited to that lactate diagnostic procedure.

# 5  Conclusion

From a mathematical point of view, this investigation shows the double phase model as the optimal fitting method for load-lactate datasets in swimming. However, for further use of the fitting curve for LT determination, a single curve is often needed. In this case the exponential function can be confirmed as the default method.

# References

1. Association, G.S.: Komplexe leistungsdiagnostik testbeschreibung. Schwimmen Lernen und Optimieren 17, 168 (2000)
2. Beaver, W.L., Wasserman, K., Whipp, B.J.: Improved detection of lactate threshold during exercise using a log-log transformation. Journal of Applied Physiology 59, 1936–1940 (1985)
3. Bosquet, L., Leger, L., Legros, P.: Methods to determine aerobic endurance. Sports Medicine 32, 675–700 (2002)
4. Busso, T.: Variable dose-response relationship between exercise training and performance. Medicine and Science in Sports and Exercise 35, 1188–1195 (2003)
5. Cheng, B., Kuipers, H., Snyder, A.C., Keizer, H.A., Jeukendrup, A., Hesselink, M.: A new approach for the determination of ventilatory and lactate thresholds. International Journal of Sports Medicine 13, 518–522 (1992)
6. Faude, O., Kindermann, W., Meyer, T.: Lactate threshold concepts how valid are they? Sports Medicine 39, 469–490 (2009)
7. Gamma, E., Helm, R., Johnson, R., Vlissides, J.: Design Patterns Elements of Reusable Object-Oriented Software. Eddison-Wesley, Boston (2002)
8. Grant, S., McMillan, K., Newell, J., Wood, L., Keatley, S., Simpson, D., Leslie, K., Fairlie-Clark, S.: Reproducibility of the blood lactate threshold, 4 mmol center dot l(-1) marker, heart rate and ratings of perceived exertion during incremental treadmill exercise in humans. European Journal of Applied Physiology 87, 159–166 (2002)
9. Hughson, R.L., Weisiger, K.H., Swanson, G.D.: Blood lactate concentration increases as a continuous function in progressive exercise. Journal of Applied Physiology 62, 1975–1981 (1987)
10. Lundberg, M.A., Hughson, R.L., Weisiger, K.H., Jones, R.H., Swanson, G.D.: Computerized estimation of lactate threshold. Computers and Biomedical Research 19, 481–486 (1986)
11. Mader, A., Liesen, H., Heck, H.: Zur beurteilung der sportarspezifischen ausdauerleistungsfhigkeit im labor. Sportarzt Sportmedizin 27, 80–88, 109–112 (1976)
12. Pansold, B., Zinner, J.: Stellenwert der Laktatbestimmung in der Leistungsdiagnostik, chap. Die Laktat-Leistungs-Kurve - ein Analyse und Interpretationsmodell der Leistungsdiagnostik im Schwimmen, pp. 47–64. Fischer, Stuttgart (1994)
13. Rudolph, K., Berbalk, A.: Ausdauerdiagnostik im rahmen der dsv-kld von 19921997. In: Freitag, W. (ed.) Schwimmen Lernen und Optimieren. vol. 17, pp. 33–55 (2000)
14. Simon, G., Berg, A., Dickhuth, H.H., Simon-Alt, A., Keul, J.: Bestimmung der anaeroben schwelle in abhngigkeit vom alter und von der leistungsfhigkeit. Deutsche Zeitschrift fr Sportmedizin 32, 7–14 (1981)
15. Smith, D.J.: A framework for understanding the training process leading to elite performance. Sports Medicine 33, 1103–1126 (2003)
16. Snyman, J.: Practical Mathematical Optimization - An Introduction to Basic Optimization Theroy and Classical and New Gradient-Based Algorithms. Springer, New York (2005)

# Computer analysis of bobsleigh team push

Peter Dabnichki

School of Aeronautical, Mechanical and Manufacturing Engineering
RMIT, Melbourne, Victoria, Australia

**Abstract.** Bobsleigh start is a simple action requiring the crew to push as hard as possible and gain maximum initial velocity of the sled at the start. However, detailed computer analysis based on velocity and acceleration data shows that timing of loading play a very important role. In this work we demonstrate that a very important performance parameter commonly called exit velocity can be use as both target and performance measurement. The analysis of time profiles allowed us to modify the timing of the loading and gain nearly 1km/h on the exit speed.

**Keywords.** Bobsleigh, start, data structure, feedback

## 1    Introduction to the sport of bobsleigh

The sport of bobsleigh is a sport where athletic performance meets science and engineering. The roots of the sport could be dated back to the late 19th century when the Swiss attached a steering mechanism to a toboggan. It is now accepted that the informal bobsleigh competitions started in 1860. The official start as bobsleigh as a sport is in 1897 when the world's first bobsleigh club was founded in St. Moritz, Switzerland. To date Saint Moritz is considered the home of bobsleigh and it is still the only place with guaranteed competition every year and still on natural track. The sport spread quickly across the alpine countries and by 1914, bobsleigh races were taking place on a wide variety of natural ice courses. The first racing sleighs made of wood but were soon replaced by steel sleighs named bobsleighs. Bobsleigh's name became popular as crews bobbed back and forth to increase their speed on the straightaways. Te sport has become internationally recognised and governed in 1923 - the Fédération Internationale de Bobsleigh et de Tobogganing (FIBT) was founded. Immediately followed the debut of the sport in 1924 when a four-man race took place at the first ever Winter Olympics in Chamonix, France. As pretty much all other sports the true internationalisation of the sport became a reality after the Second World War. In the 1950s the professional sport had begun to take shape. There and then the critical importance of the start was recognised and strong fast athletes from other sports were drawn to contribute to improvement in performance. The improvement in both technology and performance has led to steady increase in the velocities of the sleds and as a result a number of accidents occurred with certain number of fatalities. Since the 1970s, safety has gradually become a driving issue and technical regulations specifically aiming to curb the number of casualties were introduced. In 1985 – the current regulations were drawn and have been rigorously enforced since.

   In 1985, the following sled standards were introduced by the FIBT to place all competitors on equal footing. Regarding athletic performance the rules aimed to curb both the weight of the sled and the total weight, i.e. to effectively prohibit the use of overly heavy crews that the sport has become notorious. The weight restrictions are presented in Fig.1 below.

© Springer International Publishing Switzerland 2016
P. Chung et al. (eds.), *Proceedings of the 10th International Symposium
on Computer Science in Sports (ISCSS)*, Advances in Intelligent Systems
and Computing 392, DOI 10.1007/978-3-319-24560-7_25

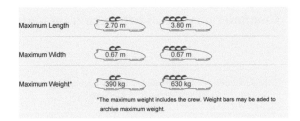

| Maximum Length | 2.70 m | 3.80 m |
| Maximum Width | 0.67 m | 0.67 m |
| Maximum Weight* | 390 kg | 630 kg |

*The maximum weight includes the crew. Weight bars may be aded to archive maximum weight.

**Fig. 1.** Bobsleigh weight and dimensions regulations for male crews

The weight of the sled itself is imposed as a minimum of 170 kg for 2-man bobsleigh and 210 kg for the 4-man bobsleigh to prevent the use of ultra-light sleds in combination with giant (sometimes overweight) athletes which brought some notoriety to the sport in the 1950s.

### 1.1 The bobsleigh start

All three Olympic track sports; luge, skeleton bobsleigh and bobsleigh use the same track. Only luge has a different start position. Both skeleton and bobsleigh adopted the so called flying start, i.e. the time is taken after crossing a line that is away from the actual starting point. The start procedure is strictly regulated and is described below. It should be mentioned that when the sport originated, crews were sitting and bobbing the sled which provided the name for the modern sport.

The crew pushes the sled from a wooden start block that must be at least 150 cm long, 20 cm wide and 5 cm high from the surface of the ice. The crew cannot pass the block or remove it prior to the start. Once the start is initiated the crew pushes the sled along the "push-off stretch" (FIBT, 2014) which is the part of the track between the start block (wooden board) and the first photoelectric cell. This stretch of the track is postulated to be 15 meters long with a gradient of 2%. As stipulated by FIBT (2014), after the first photoelectric cell (which represents the virtual start-line for the time-taking), the track must follow a straight path so that bobsleds starting off may reach a speed of 35 km/hr or 9.72 m/s.

The stretch between the first photoelectric station must be 50 m long (starting time).

**Fig. 2.** Bobsleigh 4 men start - initial push-off

After the first cell the gradient becomes very steep - 12% and all crew members need to "load" i.e. to quickly jump in the sled without any delay as the sled velocity goes quickly beyond the athletes' maximum running speed. The first person to load is the driver and brakeman is last. The order of the two pushers is decided by the team. The driver loads at the flat section usually immediately after the time start line as shown in Fig. 3 below. By that time the sled is accelerated sufficiently and the role of the driver is to steer the course. The other crew members follow the driver while maintaining a steady course notionally a straight line (in reality they need to follow the tracks created by previous starts). It should be noted that that the entire duration of the start is only about 10 s including the non-timed push-off phase. Within this short period the sled should be accelerated from stationary (status) zero velocity to reach about 10 m/s while all crew members should load the sled. From this point onwards it is down to the driver to win the race, as the rest of the crew cannot deliver any meaningful contribution apart from holding still to enable easier steering - not an easy task in view of substantial lateral loads and impacts of 10g.

**Fig. 3.** Driver loading the sled and folding the handle

## 1.2    Importance of the start

It is believed that the fastest start wins the competition which is clearly not the case but there is some correlation between the start time and the finishing of the team. There are some publications in that respect that claim to prove the point but the reality is that the track profile and weather condition affect the correlation drastically. For example Park City, Utah is considered a "starters' track" but the Olympic gold medals in 2002 were won by relatively slow starters as the race was affected by a heavy snowfall. At the same moment Igls near Innsbruck does not favor fastest starters. Regardless the opinions no one should start slower in order to win as time differences between the top six are minimal and one cannot afford to lose precious hundreds of a second. Winning crews in very high percentage tend to be among the top six starters. Hence, teams spend substantial portion of their time off-season simulating a start. In order to achieve best outcome, it is important to

- select the best combination of athletes and
- as a result of the above to achieve optimal synergy.

Frequently sprinters are attracted to the sport as they possess natural speed. This tactics is not always productive and even sub twenty second two hundred meters sprinters were not among the fastest starters in competitions. In this work we explain how is the bobsleigh start different and how the should coaching specialists aim to improve the team performance.

## 2    Computer aided analysis of the bobsleigh start

In order to improve the performance one should clearly identify the aim. In the case of the bobsleigh start the aim is frequently misconceived as to run the fastest time between the -15 and 50 meters. The real aim is to achieve maximum velocity at the end of the run and is important to underline velocity rather than speed.

In order to achieve this one needs to gain the optimum performance from the team as a whole rather than the individuals. It is important to note that the speed of the sled cannot exceed the speed of the slowest team member at any point of time. As the brakeman is the last man loading, he needs to be the fastest at he loads when the sled's speed is increasing fast, normally at the beginning of a steep slope. Ideally at this point we found that the brakeman could add further momentum at

## 2.1    Performance parameters

In order to increase the momentum of the object the crew accelerates it by applying a push force. The desired outcome is maximum momentum  in most effective manner by all members of the crew. This is a collective effort and the individual contribution is not always clear. In order to develop a performance model we needed to define the parameters. For ease of understanding and to keep the work short we focus on two-man start routine which involves less parameters.

There we have basically the following simple routine

1. Joint non-timed push-off, i.e. between the -15 to 0 the start gate
2. Driver loading
3. Brakeman push phase
4. Brakeman loading.

**Fig. 4.** Two man start

We reiterate that the aim of the bob start is to generate optimum velocity profile ending with a maximum velocity in optimal direction. This means to generate an optimal velocity profile rather than to cover the distance in minimum time.

This statement is somewhat counterintuitive to the athletes' instincts but the competition just begins with the start and does not end with it.

There are two main differences with plain sprinting are:

1.  Athletes need to accelerate the sled (impart momentum to the object) and for a two–man crew this is equivalent to sprint start when person's own weight is doubled. In order to form a creww individual tests are conducted at the end of the strength and conditioning cycle.
2.  The measured quantity is effectively group time; hence athletes need to run with equal speed, effectively the speed of the slowest person. The purpose of the technique training is to build this speed up in controlled fashion so maximum exit velocity is achieved.

Hence the analysis ought to be conducted at two levels: system level (the system is represented by the sled and the system level measurements are its velocity and acceleration) and individual level - the imparted momentum of the individual athlete on the sled. We then established that the system performance is purely measured by the kinematic parameters of the sled - i.e. velocity and acceleration and the individual contribution in proportional to the force applied in the direction of the track (i.e. the component parallel to the track) which we called the push component. However is should be noted that there is also a negative contribution by the athletes, the downward component of the force and the lateral components (both leading to increase friction on increased friction between ice and runners).

It is now obvious what data are related to the assessment of the performance and how should they be obtained. The proposed solution was two-fold

1. Obtain velocity/acceleration profile for the run – we used accelerometer in the sled, and
2. Assess qualitatively the force imparted by the athletes on the sled (i.e. their relative rather than the absolute contribution).

It was not possible to built a highly accurate system as no electronic devices could be mounted on the sled permanently. Further to this on official training runs no such devices should be used. Hence a easily to mount and remove system that is still safely mounted .At the time only simple time gaits were used and they did very little to assess the individual contribution as well as the exit velocity could not be obtained. In order to obtain those we needed were data that cover the entire duration of the start including the loading and we instrumented the sled as shown in Fig.5 below. We needed two transducers on the handles at the back of the sled where the brakeman grips and one at the drivers folding handle. We tried but could not use accurate load cells as the mounting requirements could not be met hence we used adhesive strips with effectively strain gauge system. The time pressure and variability of the athletes grip made it impossible to accurately calibrate such system. Based on the measurements we assessed the timed contributions and matched them with the acceleration and velocity profiles of the sled. The data were also used to provide formative feedback to the athletes to help them time and synchronize their actions. We found that there is not a superior technique in the simple case of the brakeman initiating the push and then the driver "hitting" the sled flying versus both of them starting the push. However, if there is a mismatch in strength and speed between the driver and the brakeman (i.e driver is slower and less explosive), then the driver hitting the sled and loading a stride earlier is a better option. It is not unusual for the drivers to be less athletic as a number of them develop later and win medals till their late thirties.

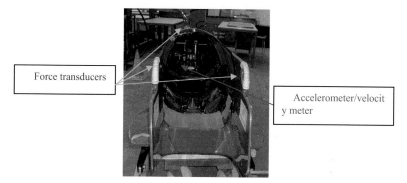

**Fig. 5.** Positioning of the force transducers and accelerometer/velocity meter

## 2.2    Sled acceleration and velocity

We here illustrate one simple example of what the acceleration measurement provides. Supposed that the standard timing gates were just used we shall not be able to see any of these developments . But why are we measuring both velocity and acceleration. The answer is simple - the velocity is built up gradually as the acceleration is produced entirely by the athletes' push on the flat part. However when the slope is reached the athletes contribution dwindles very quickly and they need to "load" the sled prior to their contribution becoming negative. We show the velocity and acceleration profiles in Fig. 6 below. One very important performance parameter is the timing of the load. This is utterly important when there is a mismatch – i.e. the driver is less athletic than the brakeman. In the case illustrated below the driver is a stride late and starts "pulling back the sled". As he struggles to enter the brakeman slows down and then accelerates the sled. The exit speed reached was 34.89 km/h, well below the target speed of 35.6 km/h. The latter was achieved by cutting the driver run by a single stride.

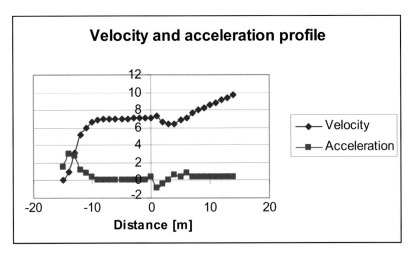

**Fig. 6.** Force pattern in time - light blue indicates brakeman's contribution

This approach allowed faster feedback (immediately after completion of the start due to very little need for calibration and filtering), quick processing and easy to communicate the data to the athletes. In effect it proved unnecessary to measure forces whose magnitude is not reliably obtained. The forces' magnitudes are related to bending moments of asymmetric beams and arcs that are highly nonlinear and very sensitive to centre of pressure fluctuation. Such fluctuations are inevitable as athletes have different trajectory than the sled and their grip force changes direction. This in effect helped us build the dual level model that focuses on system performance - i.e. velocity profile and acceleration pattern. This is sufficient for two man crew while complex synergies in the four man sled cannot be extracted from these relatively simple profiles. This allowed us to not only look into the effect on those characteristics but also define target and real performance. Below is shown again the acceleration pattern following the modified push when the driver loading into the sled, the reduction in acceleration is still evident and the graph looks surprisingly the same but the exit velocity was 35.54 km/h leading to a better overall time.

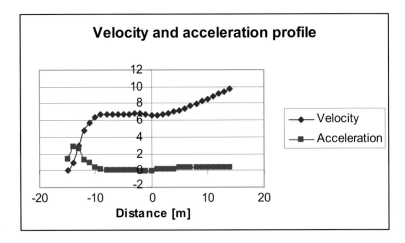

**Fig. 7.** Velocity and acceleration profile with earlier

## 3     Discussion and conclusions

### 3.1     Comparison to existing developments

Bobsleigh is predominantly perceived as technology sport as underlined by the fact that all bobsleigh publications in reputable journals are related to technology aspect (see Braghin et al (2011), Dabnichki and Avital (2006) and Dabnichki et al (2004)) while the athletes contribution is somewhat left behind. There are two notable exceptions in relation to the training of athletes. Kelly and Hubbard (2000) reported the first simulator built to train the US bobsleigh drivers. More recently Lembert et al (2011) described a system for feedback provision on a luge start. Both papers are focused on the technological development of a training aid and the way and very little attention is paid to the actual performance parameters - these aspects are left to the coaches. Further to this both systems are devised to train individuals and have no facility for team synergy development which is a common feature to most if not all simulators and feedback systems.

### 3.2     Our approach

We decided to break with the tradition of building elaborate biomechanical models and attempt to exactly measure and assess each individual parameter. We simply mapped the individual contribution on the system performance level and established the important characteristics that affect performance. This allowed us to focus the attention of both coaches and athletes on how to plan, train and execute the elements of the start. However, having a clear performance target that clearly affects the standing of the crew – i.e. exit velocity we were able not only to demonstrate but introduce appropriate modifications. Here we should mention that unlike traditional research sport performance cannot afford to spend time building statistically significant data sets as the time is limited and controlled tests virtually impossible. Once we broke with the tradition of centering the system on measurement we have discovered that the so perceived "noise" in the signal provides us with valuable information on the interaction between the athletes and the sled - hence we were able to extract the mismatch in timing of the athletes' strides. We deliberately avoided the illustration of force patterns as they are clearly noisy but important events were captured and used to provide evidence to the athletes. This was clearly the case to convince the driver to load earlier as he was convinced that the slow down was caused by the brakeman. In the more complex case of four man crew we found demonstrated that stride length and frequency should match in order to build a smooth acceleration profile and use less energy. We found that weather conditions dictate the optimal stride length and on faster tracks athlete can use longer strides but this increase the risk of loading accidents. In general we found that a strong brakeman could compensate for a less athletic s driver and redesigned driver's contribution limiting it to the very initial push off. We also used simple tools to generate the synergy and reduce the "twisting" moment that the driver tends to apply on the sled. We used statisticians but their elaborate multi-parametric models were of very little use and this is a common problem for elite sport studies as the samples are both small and inconsistent. Further to this there are factors such as ice hardness, sun light, air and ice temperature, track state that need to be assessed on the day and addressed in the target setting – exit velocity is the best example.

### 3.3     Further developments in the data analysis

Almost by accident we matching the velocity profiles to the distance and then added acceleration. At the moment we consider this approach could fully replace time distributed data. However, it is slightly early to speculate as the theoretical apparatus require formal expression of the velocity profile that needs to be related to the actual terrain and conditions and this is not an easy task. Still as evident from figures 6&7 we still gain quite valuable insight and could assess the effectiveness of decisions taken.

### 3.4     Future directions

In addition to the development in the phase space analysis we need to look into ways to formalise the data analysis and look relate the findings to explicit recommendations for performance alterations which includes finding optimal match of athletes within a given crew, timing and sequence of loading for four crew members. There is also a need to relate motion analysis, specifically stride pattern and momentum imparted on the sled. There is a wide spread problem of athletes pushing the sled to far forwards and then effectively holding it back. There are also more to be gained from understanding the power flow and dampening of adverse forces achieved by the athletes during the push-off.

## References

F. Braghin, F. Cheli, M. Donzelli, S. Melzi, E. Sabbioni (2011) Multi-body model of a bobsleigh: comparison with experimental data. *Multibody System Dynamics*, **25**, 185–201.

P. Dabnichki and E. Avital. (2006) Influence of the postion of crew members on aerodynamics performance of two-man bobsleigh. *Journal of Biomechanics*, **39**, 2733-2742.

P. Dabnichki, F. Motallebi and E. Avital (2004) Advanced bobsleigh design. Part 1: Body protection, injury prevention and performance improvement. *Proceedings of The Institution of Mechanical Engineers, Part L: Journal of Materials: Design and Applications,* **218**, 129-137.

FIBT (2014) International Bobsleigh Rules 2014. (available at FIBT.com)

A. Kelly, M. Hubbard (2000) Design and construction of a bobsled driver training simulator. *Sports Engineering*, **3**, 13-24

S. Lembert, O. Schachner and C. Raschner (2011) Development of a measurement and feedback training tool for the arm strokes of high-performance luge athletes. *Journal of Sports Sciences*, **29**, 1593–1601.

# Part V
# Virtual Reality

# Development of a Novel Immersive Interactive Virtual Reality Cricket Simulator for Cricket Batting

Aishwar Dhawan[1], Alan Cummins[1], Wayne Spratford[2], Joost C. Dessing[1] and Cathy Craig[1].

[1]School of Psychology, Queen's University of Belfast, 18-30 Malone Road, Belfast BT9 5BN
[2]Movement Science, Australian Institute of Sport, Canberra, Australia

**Abstract.** Research in the field of sports performance is constantly developing new technology to help extract meaningful data to aid in understanding in a multitude of areas such as improving technical or motor performance. Video playback has previously been extensively used for exploring anticipatory behaviour. However, when using such systems, perception is not active. This loses key information that only emerges from the dynamics of the action unfolding over time and the active perception of the observer. Virtual reality (VR) may be used to overcome such issues. This paper presents the architecture and initial implementation of a novel VR cricket simulator, utilising state of the art motion capture technology (21 Vicon cameras capturing kinematic profile of elite bowlers) and emerging VR technology (Intersense IS-900 tracking combined with Qualisys Motion capture cameras with visual display via Sony Head Mounted Display HMZ-T1), applied in a cricket scenario to examine varying components of decision and action for cricket batters. This provided an experience with a high level of presence allowing for a real-time egocentric view-point to be presented to participants. Cyclical user-testing was carried out, utilisng both qualitative and quantitative approaches, with users reporting a positive experience in use of the system.

**Keywords:** Virtual-reality; Presence; Cricket; Sport Psychology; Training; Performance.

## 1    INTRODUCTION

Scientific understanding of sports performance has a multitude of advantages; such as devising efficient training protocols, optimising decision making skills and reducing injuries. To this end, scientists and coaches are constantly adopting new and emerging technologies to extract meaningful information about the true nature of an athletic performance. However, the examination of an athlete's performance in a controlled environment still remains a challenging task. Furthermore, to obtain an in-depth understanding of expert decision-making in sport, it is essential that the correct methodologies, which allow reliable recreation of the perception – action cycle, are used. In

© Springer International Publishing Switzerland 2016
P. Chung et al. (eds.), *Proceedings of the 10th International Symposium on Computer Science in Sports (ISCSS)*, Advances in Intelligent Systems and Computing 392, DOI 10.1007/978-3-319-24560-7_26

addition, a technology that allows for the recreation of an athlete's three-dimensional viewpoint (referred as an ego centric viewpoint) of an unfolding sports related event, updated in real time, is essential. The most extensively adopted tool for exploring anticipatory behaviour to date has been video playback. This methodology is often applied because of its relative low cost, accessibility and ease of capture. However, it is limited as the viewpoint presented during an experimental scenario is fixed to that of the camera during the recording. In other words, the perception is not active. This prevents exploring data from key variables that emerge from the dynamics of the action unfolding over time and the active perception of the observer (i.e. where they choose to look). Proponents of the Gibsonian [1] approach argue the importance of the continuous role of task specific perception-action realism and its need to be maintained when investigating any motor skill. In addition, video playback is based on past events and therefore does not allow any further experimental manipulation. For instance, Shim and Carlton [2] showed a progressive deterioration in expert performance for racquet sports across "live", "film" and "point light displays". VR may be used to overcome such issues and offers an exciting solution where athletes can be immersed inside bespoke scenarios and provided with a three dimensional egocentric viewpoint. Given the rapid growth in computer processing power, limitations in rendering realistic immersive, interactive virtual environments in real time have largely been overcome. Now, athletes can interact with simulated opponents while the experimenter controls and modifies the type of information presented. Behavioural responses made by the player, including how and when they act, can also be simultaneously recorded and related back to the perceptual information. This unique approach allows us to more fully interrogate the perception-action cycle and the role of informational cues utilised by an expert player [3].

Cricket is a team sport and popularly known as the duel between "bat and ball" played on the centre of an oval shaped field on a 20.12 m rectangular area called the pitch. To score a run(s), a cricket batters is confronted with a range of bowlers (classified broadly as pace and spin) who have the ability to bowl at varied ball speeds or with different amounts of spin on the ball. For instance, a fast bowler will use swing (i.e. when the ball moves in the air whilst following its trajectory to the batter) and seam (i.e. the sideways movement, the ball trajectory undergoes once the ball bounces off the pitch and lands in front of the batter) whereas spin bowlers use lateral revolutions to put spin on the ball to cause lateral deviation off the pitch. Therefore, the batters are faced with an array of stimuli in order to make a successful interception and avoid a wicket being taken. In addition, a batter can choose from a broad range of strokes, striking the ball anywhere in their 360o arc. Research has shown the importance of representative task design in evaluating batting skill [4-5]. More recent evidence by Mann, Abernethy & Farrow [6] suggests that a batter must have the actual intent of intercepting to truly reflect performance. In lieu of this, Probatter (http://www.probatter.com) is the most advanced form of batting training equipment used [7] using a combination of video and a bowling machine to simulate a real ball trajectory. Probatter has been suggested as an ideal vehicle to further explore the perception action relationship in cricket batting [7]. However, it can be argued that having a fixed ball release point combined with the use of 2D video based information to present a three dimensional naturalistic ball trajectory reduces task realism. With this in mind, we envisaged that a VR cricket simulator could offer a new possibility of looking at cricket batting in a task specific controlled environment that still offers perception action coupling.

The following sections describe the design and implementation of such a system utilising state of the art motion capture technology and VR technology to examine varying components of decision and action for cricket batters.

## 2    DESIGN

The utilisation of VR technology is directly dependent on the complexity of the interaction between the user(s) and the environment system in order to elicit natural user behaviour and therefore requires significant understanding of the real life scenarios that are to be simulated.

### 2.1    Capturing Player Movements

Twenty-one Vicon Cameras (Oxford Metrics, Oxford, UK) were used to record the kinematic profiles of elite cricket bowlers, from a high performance programme at the Australian Institute of Sport in Canberra, Australia. They were equipped with forty-eight markers taped to specific anatomical landmarks to create a three dimensional representation of the full body set up at 250 Hz (See Fig. 1). All bowlers bowled on a piece of artificial turf, allowing a standard run-up. The motion capture volume encompassed full body capture of all bowlers from the point of back foot impact to at least 100ms post ball release. In addition to the standard trial capture, two additional trials for each bowler were captured consisting of the bowler's jump phase (the point of jump initiation to when the first frame of landing occurs) to further create a continuous run for each trial (see section 2.3). All bowlers were required to bowl to a zone

**Fig. 1.** (A) Bowler with 48 markers at key anatomical landmarks. (B, C) Motion capture volume and camera set up at AIS. (D) Sony Handicam HDR FX7 mounted above the bowler's head to estimate the delivery types. (E) View of the artificial turf (F) The calibrated area with distinct red squares for targeting aim.

indicated on a rectangular target grid All recorded motion capture data was reconstructed using the Vicon Nexus to retrieve any missing markers and filter out any
noise resulting from hidden or overlapping markers.

## 2.2    Capturing Ball Release

The initial ball flight was tracked using twenty-two Vicon Cameras. Three markers
were located on the ball to track the initial ball trajectory, out of the hand. In addition,
initial coordinates of the ball (volume allowed up to at least 5 frames of ball trajectory
post release) were used to determine the initial ball trajectory and release speed.
Moreover, a video camera, above the bowling crease, and a calibrated tape measure
on the side of the pitch was used to identify the ball bounce position and final position
on the target grid. Realistic ball trajectories were then generated using a combination
of these data and modelling. Based on the positions of three markers on the ball, the
spatial orientation of the following frames was estimated as the point that moved most
closely along a parabola/second order polynomial fit. In addition, the bounce and final
ball position was estimated using a combination of linear calibration between pixels
and centimetres and images from the video camera. Further to this, a 3D aerodynamic
model [8] that included gravity, air resistance and spin related lift forces was used to
calculate the ball trajectory. The ball trajectory parameters (spin rate, drag coefficient,
angle specifying the orientation of the rotation axis – kept constant throughout the
trajectory), and the rebound factor (reduction in ball velocity after bounce) were op-
timised to minimize the summed distances of the ball position at bounce and at the
final landing position on the grid. An additional penalty on the fit was applied to minimise the difference between the pre and post bounce ball direction in the ground
plane. Finally, the simulated ball trajectory yielding the smallest error was used in the
virtual cricket simulator and was saved at a frame rate of 100 Hz.

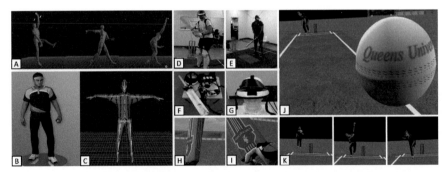

**Fig. 2.** (A) Fast bowling action with actor (B) The 3D model (C) Characterisation with Skele-
ton (D, E) Batters wearing HMD and backpack (F) Control Unit Backpack for HMD (G) Sony
HMD (H, I) Virtual cricket bat manipulated in real-time (J,K) Virtual bowler and ball visible in
HMD.

## 2.3   Tying Player Movements to a Virtual Depiction

Virtual animations were created using Motion Builder Animation Software (Motion Builder version 2012, Autodesk, California) which provided efficient motion retargeting independent of morphology. Character animation features such as a constraint solver, for instance ensuring ground contact, adapting to the position and orientation of movement using FK/IK rig (Forward Kinematics, Inverse Kinematics) were also used to create seamless character animations from the motion capture data. In addition, the cricket ball was added to the bowling hand (See Fig. 2J-K).

## 2.4   Creating the Experiment

Unity 4.0 (Unity Technologies, USA) software framework was used as a platform to simulate all experimental information for the users. This allowed:

1. Motion Scheduling: All trials were presented in a controlled manner via a properties file designating bowler animation, trial order and ball trajectory.
2. Event synchronisation: Ball release triggering a trajectory from pre-captured data.
3. Feedback generation: Auditory feedback was provided when the ball made contact with one of three rigid body cubes (representing the bottom part of the bat, upper part of the bat and edges). At the end of the experiment a scorecard outlined the number and type of contacts to provide performance feedback.
4. Environment Scaling: A virtual cricket stadium and associated objects was created conforming to a global coordinate system with a one to one mapping between virtual and real i.e. the virtual stadium replicated the exact dimensions of the real world (pitch length, crease markings, stumps, and physical dimensions of the ball).

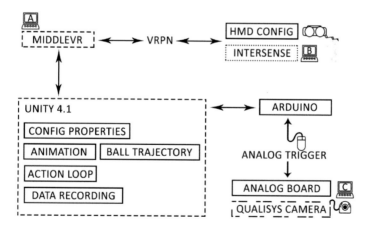

**Fig. 3.** General software / hardware architecture of the system. Comprised of three interacting computers (A) Controlled MiddleVR and the main Unity program. (B) Controlled data acquisition from the Intersense Acoustic trackers. (C) Controlled Qualisys Motion Capture Cameras and interaction with Analog data acquisition board.

## 2.5    Interaction within the Virtual Environment

User control or interactivity is considered paramount to the experience of VR as this directly relates to the level of behavioural realism experienced by the user [3], [9-10]. From a sporting performance perspective, egocentric motion tracked stereoscopic 3D images presented by Head Mounted Displays (HMDs), are advantageous as they offer a highly immersive experience, thus enhancing the behavioural realism of a task. A Sony HMZ T1 headset with a resolution of 1280X1024 and $45^{\circ}$ diagonal viewing angle was used providing a high-resolution signal over HDMI to a control unit worn by the participant (See Fig. 2F-G). This compared favourably to a Cybermind Visette 45, previously used in development of a VR rugby simulation [11]. The MiddleVR plugin was used to control stereoscopic rendering of each eye image to the end-user.

Mapping natural movement in a virtual world is judicious as it enhances the level of task behavioural presence of the user [8]. An Intersense tracker (Intersense 900 System, Massachusetts, USA) was located on the front of the headset to update the egocentric viewpoint. An additional sensor was located on a cricket bat to replicate its 6DOF real world movement inside the tracked volume of $144m^3$ (8m L X 6m W X 3m H). Use of this hybrid tracking system (using inertial and acoustic tracking) ensured minimal signal noise/drift experienced by the user within the environment. Additionally, an eight Oqus camera Qualisys motion capture system (Qualisys AB, Sweden) was used to provide system independent 3D bat movement to aid metric evaluation of performance post-experiment. Time synchronisation of these systems, within Unity, was achieved using an Arduino board, utilising the Uniduino library. MiddleVR software, using a Virtual Reality Peripheral Network stream mapped 3D data from Intersense to the Unity environment (See Fig. 3). Finally, in order to create a task representative set-up and familiarity for batters, a platform with real crease markings and stumps was constructed and corresponded to that experienced inside the virtual environment.

## 3    DESIGNING FOR PRESENCE

Presence is referred to as the subjective sense of "being" in the virtual environment [3], [12-13]. A high level of presence was achieved by:

1. Providing batters with an egocentric viewpoint updating in real time (100Hz).
2. Allowing batters to intercept the ball by physically performing cricket shots as they would normally do on a real cricket pitch.
3. Providing an accurate cricket environment. E.g. using kinematics of real bowling action synced with ball trajectories and a one-to-one mapping of world scaling.
4. Replicating 6DOF movement of a cricket bat in real time inside the virtual environment provided batters with a natural sense of control.
   Allowing batters to wear batting gloves and pads to increase levels of task engagement.

## 4    USER-TESTING

Cyclical design and user-testing was carried out to aid design. The equipment was initially designed by the researchers, all of whom were experienced in design and development of VR systems for sports. However, it was critical to incorporate the experience of the batters. Firstly, a batter from Queen's University Belfast Cricket Club was taken through a shortened version of the experimental conditions and asked:

1. If they could perceive the motion depth of the ball trajectory.
2. If they could identify the ball bounce location.
3. To attempt to hit balls in a naturalistic manner.
4. If they could perceive ball to bat contact based on sound and visual events.
5. If they felt any motion sickness or discomfort during presentation of trials.

    Overall the batter reported a positive experience. Through observation and analysing batters' feedback, the headset was optimized for comfort and reduced external light. HMDs have been criticised for constraining user's movements with the potential for control units, to interfere with performance of active tasks [14]. This was then followed by user-testing with a Cricket Ireland batter. He reported a positive experience with optimism for its use in education and training. However, he noted loss of ball trajectory at end of batting phase. This could be attributed to the batter attempting to use his gaze (batter making predictive gaze saccades is not uncommon in cricket [15]) to predict ball position rather than global head movement tracked by the sensor. Subsequently, a panel of experts from Cricket Ireland attended a demonstration session to briefly evaluate the system. Finally, a pilot experiment was conducted with a batter to investigate collision detection and whether the batter was able to maintain a high level of task engagement throughout. On multiple occasions double scores were counted whereby the ball collided with two independent collision zones. In addition, the batter did not hear sound occasionally when they thought a successful interception had occurred. This may have been due to lost collisions owing to rapid swing of the bat and ball. In order to combat this, the collision zones were increased.

## 5    CONCLUSION AND FUTURE WORK

The system described illustrates a novel immersive, interactive VR bowling simulator that may be used to assess expertise in cricket batting. However, there were a number of limitations identified relating to collision detection, equipment ergonomics, loss of data signal from trackers and field of view. Moving forward this system will be used in to evaluate a batter's ability to detect different bowling techniques. This is critical to understand what information can be used by the batters to anticipate ball delivery. Further to this the VR simulator will be utilised to test what types of shots expert and intermediate cricket batters choose to make when confronted with different deliveries.

# 6    ACKNOWLEDGEMENTS

The project was partly funded by an ERC grant awarded to the last author (ERC StIG 210007, TEMPUS_G). We would also like to acknowledge the support of Crick-et Ireland and the Australian Institute of Sport, during development of this project. The authors would also like to thank Damian Farrow, Professor of Skill Acquisition at the Institute of Sport and Active Living, Victoria University for his support and input during the project.

# 7    REFERENCES

1. Gibson, J. J. (1979) The ecological approach to visual perception. Boston; London: Houghton
2. Mifflin.Shim, J., & Carlton, L. G. (1999) Anticipation of movement outcome through perception of movement kinematics. Journal of Applied Sport Exercise Psychology 21:S98
3. Craig, C.M. (2014) Understanding perception and action in sport: How can virtual reality technology help? Sports Technology 6(4)
4. Pinder, R. A., Renshaw, I., & Davids, K. (2009) Information-movement coupling in developing cricketers under changing ecological practice constraints. Human Movement Science 28(4): 468-479
5. Weissensteiner, J., Abernethy, B., Farrow, D., & Muller, S. (2008) The development of anticipation: A cross-sectional examination of the practice experiences contributing to skill in cricket batting. Journal of Sport & Exercise Psychology 30(6):663-684
6. Mann, D. L., Abernethy, B., & Farrow, D. (2010) Action specificity increases anticipatory performance and the expert advantage in natural interceptive tasks. Acta Psychologica 135(1):17-23
7. Portus, M. R., & Farrow, D. (2011) Enhancing cricket batting skill: Implications for biomechanics and skill acquisition research and practice. Sports Biomechanics 10(4):294-305
8. Dessing, J., & Craig, C. (2010) Bending it Like Beckham: How to Visually Fool the Goalkeeper. Plos One 5(10) e13161.
9. McMenemy, K., & Ferguson, S. (2007) A hitchhiker's guide to virtual reality. Wellesley, MA: A K Peters, Ltd.
10. Brault S, Bideau B, Craig C, Kulpa R (2010) Balancing deceit and disguise: How to successfully fool the defender in a 1 vs. 1 situation in rugby. Hum Movement Science 29:412-425
11. Brault, S., Bideau, B., Kulpa, R., & Craig, C. M. (2012) Detecting deception in movement: The case of the side-step in rugby. Plos One 7(6) e37494
12. Witmer, B. G., & Singer, M. J. (1998) Measuring presence in virtual environments: A presence questionnaire. Presence: Teleoperators and Virtual Environments 7(3):225 - 240
13. Slater, M., Wilbur S. (1997) A Framework for Immersive Virtual Environments (FIVE): Speculations on the Role of Presence in Virtual Environments, Presence: Teleoperators and Virtual Environments, MIT Press 6(6):603-616
14. Miles, H. C., Pop, S. R., Watt, S. J., Lawrence, G. P., & John, N. W. (2012) A review of virtual environments for training in ball sports Computers & Graphics 36(6):714 - 726
15. Land, M. F., & McLeod, P. (2000) From eye movements to actions: How batsmen hit the ball. Nature Neuroscience 3:1340-1345

# Multi-Level Analysis of Motor Actions as a Basis for Effective Coaching in Virtual Reality

Felix Hülsmann, Cornelia Frank, Thomas Schack, Stefan Kopp, Mario Botsch

Cluster of Excellence Cognitive Interaction Technology (CITEC)
Bielefeld University

**Abstract.** In order to effectively support motor learning in Virtual Reality, real-time analysis of motor actions performed by the athlete is essential. Most recent work in this area rather focuses on feedback strategies, and not primarily on systematic analysis of the motor action to be learnt. Aiming at a high-level understanding of the performed motor action, we introduce a two-level approach. On the one hand, we focus on a hierarchical motor performance analysis performed online in a VR environment. On the other hand, we introduce an analysis of cognitive representation as a complement for a thorough analysis of motor action.

## 1 Introduction

In recent years, several approaches to Virtual Reality (VR) motor learning have been presented. For instance Rector et al. [3] introduced a virtual yoga coach for visually impaired people. The authors followed a rule-based approach to define desired yoga poses. Rules were extracted from literature and interviews with experts. However, only desired optimal poses were specified, which did not take into account a hierarchical representation of motion: only simple features (joint angles calculated from joint positions) were considered. To analyze feedback strategies, Sigrist et al. [6] conducted an experiment investigating the impact of haptic, auditive, and visual feedback on motor learning using a VR rowing simulator. The simulator provides cues on performance, depending on temporal and spatial deviations of the optimal oar blade movement. Here, only the resulting motion of the oar is analyzed, not the actual motor performance of the athlete. Another coaching system was developed by Tang et al. [7] with a focus on dance training: Their analysis uses a block matching algorithm to compare the athlete's motion — in terms of joint angles — to pre-recorded optimal performances. Thus, a comparison of online movement and the desired optimal movement is possible without having to define any rule by hand. However, the system does not respect multi-layer representations, e.g., typical error patterns: It only determines which type of dance movement is executed and then calculates an accumulated error value for the performance. No fine grained analysis and interpretation of non-optimal parts of the performance is done.

Overall, these approaches perform analysis of motor actions with respect to only a subset of particular aspects of motor actions: Often only simple features

© Springer International Publishing Switzerland 2016
P. Chung et al. (eds.), *Proceedings of the 10th International Symposium on Computer Science in Sports (ISCSS)*, Advances in Intelligent Systems and Computing 392, DOI 10.1007/978-3-319-24560-7_27

of the motion (e.g., joint angles) are considered and other important features like speed etc. cannot be integrated into the model. As a detailed analysis is helpful for further steps, e.g., giving helpful feedback, this is one of the gaps we aim to diminish in this work: We develop a suitable representation and analysis of (a) online motor performance, but also with respect to (b) an offline aquisition of the athlete's cognitive representation of the of the action as presented in [4]. In our scenario, an athlete, who is tracked by an OptiTrack Prime13W system, is placed inside a two-sided CAVE (stereo projection area on floor and front). Here, she has to perform a motor action (e.g., squats) in front of a virtual mirror. The mirror reflects the athlete's motion mapped on a virtual avatar. As exemplary actions we use squat performances. The approaches discussed in the following can also be extended to further types of motor actions.

## 2   Methodology and Realization

*Analysis of Motor Performance:* Our real-time performance analysis is able to combine an extendable set of features into a hierarchical representation. It is based on rules that describe the desired motion. This kind of analysis is highly efficient and allows a direct interpretation of the results in terms of performance flaws. Motion tracked by the OptiTrack system is transferred to our framework, where we represent it as a temporal sequence of a set of features (this could be, e.g., joint angles, positions, symmetry etc.). These features can be calculated using the raw motion capture data. The sequence is split into single repetitions of motor actions (e.g., squats), connected by arbitrary transition movements. Each repetition is a combination of simpler sub-actions, denoted *Movement Primitives* (MP). This could be, e.g., the going-down and the going-up stages during a squat. Also, additional MPs, like an is-down stage, can be defined if required for the performance analysis. For each type of action and MP, a list of relevant features is manually specified. Then, key-postures for the MPs are defined using manual analysis of recorded video and/or tracking data. For the squat, this can be performed — among others — via observing symmetric key angles of knees. To detect a single action, the system has to detect a posture similar enough to one of these key-postures. The analyzer is a state machine: As soon as a posture similar enough to a key-posture describing a valid state transition is detected, the analyzer switches its state and waits for the next key-posture. If the observed motion frame does not belong to the current state nor to an allowed state transition, the system returns to the starting state: The performed movement does not depict any known action/MP or the performance has been aborted. The state of the analyzer reflects the current action and MP (see Figure 1). This representation has the advantage to allow focusing on style patterns on an extensible feature set which are especially relevant for single parts of the action.

  To detect erroneous performances, a list of Prototypical Style Patterns (PSP) is defined for every action, describing movement styles considered as undesired. A PSP is defined using at least one rule, which describes, e.g., the violation

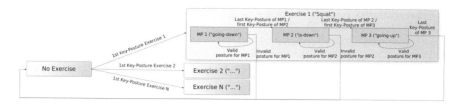

**Fig. 1.** State machine used to determine current action and Movement Primitive.

**Fig. 2.** The athlete is in the Is-Down-Stage. Here, he did not yet reach optimal depth (marked in red). During going down, PSP "Back Straight" was violated.

of specified feature constraints. Each rule returns a quantitative error value for the incoming motion. PSPs and rules are developed based on literature and information received from experts. Here, also observations from recorded data of correctly or incorrectly performed actions were taken into account. One example for a PSP in the context of squat is, e.g., "Incorrect Weight Distribution": One indicator for this pattern is that the knees are in front of the toes. Furthermore, this pattern can be detected via observing the angle of the shin or the ankle. Also using center of mass of upper and lower body may be conceivable. Finally, the highest level of this hierarchical motion representation is the action. Lower levels are the MPs and PSPs, and the lowest level are the features itself, e.g., joint angles, speed or positions at a given time. For a typical setup, our analysis needs approximately 1ms per motion frame. Figure 2 shows the system in action.

*Analysis of Cognitive Representation:* Analysis of cognitive representation is performed using structural-dimensional analysis (SDA-M) including a splitting task of a given set of basic action concepts (BACs), a hierarchical cluster analysis, and an analysis of invariance [4]. As such, this method provides psychometric data regarding the relations and groupings of BACs of a complex action (here: the squat) in long-term memory. For instance, the BACs "straight posture" and "feet shoulder-width" functionally relate to the preparatory phase of the squat, while the BACs "move bottom backwards" and "bend legs" functionally relate to the main phase. By determining the degree of BAC order formation in memory (i.e., relation of BACs in terms of functional movement phases), we can derive information both on the skill level of the athlete (cf. [5]) as well as on the athlete's progress of learning (cf. [1]). In addition, errors can be inferred by comparing an athlete's representation to a reference, e.g., an expert representation.

## 3   Discussion and Perspective

This paper presented a multi-level analysis of motor actions as a basis for an intelligent coaching space in the field of motor learning. We started with our online analysis of motor performance followed by an offline analysis of the athlete's cognitive representation. The online analysis of motor performance is highly efficient and easily interpretable. While Rector et al. [3] already successfully applied a rule-based approach to detect desired joint angles, our approach goes beyond: We respect the hierarchical properties of motion and allow considering a nearly arbitrary and extendable set of features. Thus we do not only wait for desired angles and return the current deviation, but we provide an online interpretation of performed erroneous motion in terms of style pattern. The analysis of the cognitive representation serves as a second source of information in order to learn about the individuals perceptual cognitive prerequisites for subsequent feedback and coaching strategies. As such, the cognitive analysis adds an additional high-level layer to the hierarchy, built-on by the online analysis of motor performance. In addition, coupling PSPs from motor performance analysis with groupings and relations of BACs from the analysis of the athlete's cognitive representation allows to link the level of cognitive representation to the level of motor performance [2]. Our next steps aim at investigating the potential of this combination as a basis for VR coaching.

**Acknowledgments:** This work was supported by the Cluster of Excellence Cognitive Interaction Technology 'CITEC' (EXC 277) at Bielefeld University, which is funded by the German Research Foundation (DFG).

## References

1. Frank, C., Land, W.M., Schack, T.: Mental representation and learning: the influence of practice on the development of mental representation structure in complex action. Psychology of Sport and Exercise 14(3), 353–361 (2013)
2. Land, W.M., Volchenkov, D., Bläsing, B.E., Schack, T.: From action representation to action execution: exploring the links between cognitive and biomechanical levels of motor control. Frontiers in Computational Neuroscience 7, 1–14 (2013)
3. Rector, K., Bennett, C.L., Kientz, J.A.: Eyes-free yoga: an exergame using depth cameras for blind & low vision exercise. In: Proc. of the 15th Int. ACM SIGACCESS Conf. on Computers and Accessibility. p. 12. ACM (2013)
4. Schack, T.: Measuring mental representations. Measurement in sport and exercise psychology, Human Kinetics (2012)
5. Schack, T., Mechsner, F.: Representation of motor skills in human long-term memory. Neuroscience letters 391(3), 77–81 (2006)
6. Sigrist, R., Rauter, G., Marchal-Crespo, L., Riener, R., Wolf, P.: Sonification and haptic feedback in addition to visual feedback enhances complex motor task learning. Experimental brain research 233(3), 909–925 (2015)
7. Tang, J.K., Chan, J.C., Leung, H.: Interactive dancing game with real-time recognition of continuous dance moves from 3d human motion capture. In: Proc. of Int. Conf. on Ubiquitous Information Management and Communication. p. 50 (2011)

# Part VI
# Sensing Technology

# Evaluating the Indoor Football Tracking Accuracy of a Radio-Based Real-Time Locating System

Thomas Seidl[1], Matthias Völker[1], Nicolas Witt[1], Dino Poimann[2], Titus Czyz[2], Norbert Franke[1] and Matthias Lochmann[2]

[1] Fraunhofer Institute for Integrated Circuits, Nordostpark 93,
90411 Nuremberg, Germany
[2] Friedrich-Alexander-University Erlangen-Nuremberg, Gebbertstr. 123b,
91058 Erlangen, Germany

**Abstract.** Nowadays, many tracking systems in football provide positional data of players but only a few systems provide reliable data of the ball. The tracking quality of many available systems suffers from high ball velocities up to 120km/h and from the occlusion of both the players and the ball.

Radio-based local positioning systems use sensors integrated in the ball and located on the players' back or near the shoes to avoid such issues. However, a qualitative evaluation of the tracking precision of radio-based systems is often not available and to the best of our knowledge there are actually no studies that deal with the positional accuracy of ball tracking.

In this paper we close this gap and use the RedFIR radio-based locating system together with a ball shooting machine to repeatedly simulate realistic situations with different velocities in an indoor environment. We compare the derived positions from high speed camera footage to the positions provided by the RedFIR system by means of root mean square error (RMSE) and Bland-Altman analysis.

We found an overall positional RMSE of 12.5cm for different ball velocities ranging from 45km/h to 61km/h. There was a systematic bias of 11.5cm between positions obtained by RedFIR and positions obtained by the high speed camera. Bland-Altman analysis showed 95% limits of agreement of [ 21.1cm, 1.9cm]. Taking the ball diameter of 22cm into account these results indicate that RedFIR is a valid tool for kinematic, tactical and time-motion analysis of ball movements in football.

## 1 Introduction

Positional data of football player movements helps to analyze the players' physiological demands during matches, to analyze tactical movements of opponents, and to show additional information about the performance of players to spectators. Nowadays, there are different tracking systems available that provide positional data of players. In official matches camera-based systems are used frequently as rules do not yet permit GPS- and radio-based systems that need to

© Springer International Publishing Switzerland 2016
P. Chung et al. (eds.), *Proceedings of the 10th International Symposium on Computer Science in Sports (ISCSS)*, Advances in Intelligent Systems and Computing 392, DOI 10.1007/978-3-319-24560-7_28

integrate sensors into the ball or to attach them to players. Hence, sensor-based systems are more common in training environments [18].

However, the tracking performance of camera-based systems suffers from considerable shortcomings: ball velocities of up to 120km/h, instantaneous movements, changing weather and illumination conditions and the occlusion of both the players and the ball are common challenges for these systems [14].

Several researchers have tried to evaluate the performance of different player tracking systems. See [2] for an overview how player tracking data has been used in research in the last few years. Positional data forms the basis for further statistical analyses, e.g. covered distances, runs with different intensities, analysis of tactical patterns. Thus the evaluation of the accuracy of positional data should be an integral part of the evaluation of a tracking system.

However, there are only a few studies that evaluated the positional accuracy of tracking systems [16, 18] rather than assessing the quality of a system by evaluating parameters directly, that are typically derived from positional data, e.g. covered distances and mean velocities [6,8]. This is mainly imposed by the lack of reference systems that precisely determine the position of fast moving objects. These studies have in common that they are limited to player tracking as no study tested the positional accuracy of ball tracking so far.

Although ball tracking in sport is a vivid research area within Computer Vision (Football [13] , Baseball [11], Tennis [17], Basketball [3], Volleyball [4, 9]) the performance of tracking algorithms is typically measured by means of identification rates or pixel differences whereas resulting differences in $2D$ or $3D$ positions to a gold standard should be considered.

Kelley et al. validated an automated ball velocity and spin rate estimator that works on images from a high speed camera and compared it to velocities found with the help of light gates [12].

However, to find a correct estimate for the position of fast moving objects is significantly more challenging. Choppin et al. provided a set-up for obtaining precise three-dimensional positions of fast moving objects using two synchronized high speed cameras that has been applied in tennis matches for analyzing ball and racket speeds [5].

We use a similar approach based on one high speed camera (HSC) to provide ground truth values for the position of football shots that were simultaneously tracked by the RedFIR system. The RedFIR radio-based local positioning system uses sensors integrated in the ball and located near the players' shoes to provide positional, velocity and acceleration data on the players and the ball.

We organized the remainder of the paper as follows: Section 2 explains the functional principle of the RedFIR Real-time Locating System used for the experiments we describe in Section 3. We present results in Section 4, provide a discussion in Section 5 and summarize our conclusions in Section 6.

## 2   The RedFIR Real-Time Locating System

The RedFIR Real-Time Locating System (RTLS) is based on time-of-flight measurements, where small transmitter integrated circuits emit burst signals. Antennas around the pitch receive these signals and send them to a centralized
unit which processes them and extracts time of arrival (ToA) values. ToA values are the basis for time difference of arrival (TDoA) values, from which x, y,
and z coordinates, three-dimensional velocity and acceleration are derived using
hyperbolic triangulation.

The RedFIR system operates in the globally license-free ISM (industrial, scientific, and medical) band of 2.4GHz and uses the available bandwidth of around
80MHz. Miniaturized transmitters generate short broadband signal bursts together with identification sequences. The locating system is able to receive an
overall of 50,000 of those signal bursts per second. The installation provides 12
antennas that receive signals from up to 144 different transmitters. Balls emit
around 2,000 tracking bursts per second whereas the remaining transmitters
(61mm × 38mm × 7mm) emit around 200 tracking bursts per second. The minia-
ture transmitters themselves are splash-proof (in case of the player transmitters)
or integrated into the football. Figure 1b shows a glass model of a ball transmitter. For a more detailed description of the RedFIR system and the generated
data streams see von der Grün et al. [10] and Mutschler et al. [15].

## 3   Methods

### 3.1   Hardware Setup

We conducted our experiments in the Fraunhofer Test and Application Center
L.I.N.K. in Nuremberg, where -within an area of 30m×20m×10m- a RedFIR system (version 1.1) is installed [7]. We placed a Seattle Sport Sciences SideKick ball
shooting machine at a distance of 5.5m from a target wall and shot thirty times
with speed levels 3 (4, 5), i.e., with approximately 45km/h (53km/h, 61km/h).
For better readability we will refer to these velocities as 'slow', 'medium' and
'fast'. We only activated the ball's transmitter and two reference transmitter to
minimize biasing side effects. Twelve receivers were active during our tests.

To map the coordinates of the ball to its real coordinates we used a Weinberger
G2 high speed camera with a resolution of 1536 × 1024 @ 1,000fps and a shutter
time of 992$\mu$s. The camera was adjusted to aim at the target wall, as shown in
figure 1a. Additional flicker-free light ensured that distortion or blurring effects
in the images were avoided. The camera was triggered as soon as the ball became
visible in the images of our reference system.

To calibrate the camera we used multiple checkerboard images together
with the freely available camera calibration and digitization software Check2D
(www.check2d.co.uk) and manually marked the ball in the images (Reprojection
error 0.112 pixel). As a result we obtained real world coordinates of the ball in
the direction of motion for each image frame.

We investigated the accuracy of the high speed camera data by digitizing known

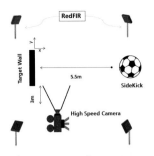

(a) Hardware setup for test measurements.

(b) A glass model of the ball's transmitter and inductive charger.

Fig. 1: Test setup and glass model of a ball transmitter.

coordinates of a grid painted on a panel (residuals: 0.5mm ± 0.2mm) and by placing the ball at known positions (residuals: 11mm ± 6.8mm) in front of the camera. As we assume RedFIR's position errors to be a magnitude higher this suffices our requirements.

### 3.2 Synchronization

To synchronize the data of the RedFIR system with the camera data we applied the following procedure:
RedFIR provides approximately twice as much positions compared to the high speed camera recordings (2000 per second). We identify the frame that shows the ball deflecting by the target wall in the high speed camera images and in the RedFIR data. Since we know the time period of the ball being visible in the camera images we can cut the corresponding RedFIR data around the identified moment of deflection. We then interpolate RedFIR and high speed camera data to 2000Hz and correlate that position with the current frame provided by the camera.
In order to specify a common coordinate system for the high speed camera data we used a grid printed on a panel in line with known coordinates in the RedFIR coordinate system. The axes of the high speed camera coordinate system were chosen to point parallel to the RedFIR coordinate system. Hence, we can transform the data points with a simple translation (and mirroring of the axes), and align the data of the high speed camera to the RedFIR coordinate system and vice versa.

### 3.3 Data Analysis

We analyzed thirty shots at three different velocities (ten trials each). To minimize biasing effects we restricted our analyses to one sequence before impact with the target wall. The sequence starts when the ball becomes visible in the image and ends 2.5ms before impact. Due to the deformation of the ball at impact it is difficult to mark the ball by fitting a circle around it.

We then calculated differences between the positions of the RedFIR system and the positions provided by the high speed camera and summarized them by means of root-mean-square error (RMSE) and 95% limits of agreement (LOA) for the shot in moving direction.

The RMSE is a measure of the deviation between the RedFIR data and the data provided by the camera and is defined as:

$$\text{RMSE} = \sqrt{\frac{1}{N} \sum_{i}^{N} (X_{rf}^{i} - X_{hsc}^{i})^2}, \qquad (1)$$

where $X_{rf}^{i}$ and $X_{hsc}^{i}$ denote the $i$-th sample, i.e. the position provided by RedFIR and the high speed camera. $N$ equals the total number of samples.

## 4   Results

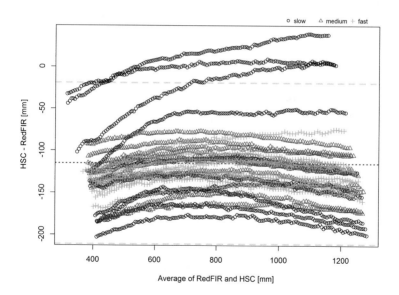

Fig. 2: Bland-Altman plot: The x-axis corresponds to the mean of RedFIR and HSC position in the direction of movement, whereas the y-axis shows the difference between the two systems. The dotted and dashed lines are the mean of the differences and the 95% limits of agreement, respectively. Circles, triangles and pluses correspond to slow, medium and fast velocities.

We were able to analyze all thirty shots. The RedFIR system provided continuously data throughout the experiments. There were no outliers in the data and we ended up comparing 3614 positions. The mean duration of the measurement interval was 0.065s.

For the comparison of the two systems we used the method by Bland and Altman [1]. Figure 2 shows the corresponding Bland-Altman plot.

Ball positions provided by RedFIR showed a systematic bias of −11.5cm. However, the overall standard deviation was 4.9cm and therefore quite low. The lower and upper 95% limits of agreement were −21.1cm and −1.9cm. The results show only a 1mm difference in RMS errors between slow, medium and fast shots. The maximum deviation between the two systems was found at lowest speed with an error of 20.3cm. The correlation between the positions obtained by the two systems was 98.1%. Table 1 summarizes the results for the different velocities.

Table 1: Positional errors obtained for different velocities: mean, standard deviation, RMSE, 95% LOA and maximum error are shown in cm.

| Velocity | $\mu$(cm) | $\sigma$(cm) | RMSE | 95%-CI | Max. error(cm) |
|---|---|---|---|---|---|
| slow | -10.2 | 7.1 | 12.4 | [-24.1, 3.7] | 20.3 |
| medium | -12.2 | 2.6 | 12.5 | [-17.2, -7.0] | 18.5 |
| fast | -12.4 | 1.8 | 12.5 | [-15.9, -9.0] | 16.8 |
| ∅ | -11.5 | 4.9 | 12.5 | [-21.1, -1.9] | 20.3 |

## 5   Discussion

Our studies show that the RedFIR system is able to accurately track the position of a football.

Its positional accuracy is better than the ones reported in previous studies for player tracking by Siegle et al. [18] and Ogris et al. [16].

For applications of positional data for kinematic and tactical analyses the estimation error of tracking systems should be below the diameter of the human's body when dealing with player data. Considering a ball diameter of 22cm the results indicate that the system is applicable in these domains for the analysis of ball movement. However, the system is not suited for applications like goal detection where only a maximal error of 1.5cm is allowed.

By using only one camera, we have limited our study to only focus on the main direction of motion. By using two synchronized cameras we could measure the tracking precision in much greater detail. However, the ball does not move much in y-direction and a comparison of the accuracy in x- and y-direction at the same time results in imprecise conclusions. Instead, we propose to rotate the set-up by 90 degrees to investigate the system's accuracy in y-direction separately. The errors are expected to be similar to the errors described in this paper. The tracking accuracy was higher for lower velocities. RMS errors for the tested range of velocities were similar. Small differences between trials prove the robustness of the RedFIR ball tracking.

The experimental design may have had an influence on the results as the position of the integrated chip, suspended in the middle of the ball, is affected by its deformation when the ball hits the target wall. The moment of deflection identified in the images corresponds to the frame when the ball visually changes its direction whereas the moment identified in the RedFIR data corresponds to the moment when the integrated chip changes its direction. These estimates do not have to agree perfectly and this could have led to an imperfect synchronization. The setup was chosen to minimize biasing effects for the high speed camera and the RedFIR system. The camera system is able to provide a very accurate estimate of the ball position for a small volume in front of the target wall and a short time interval (1m and 0.15$s$ in this setup), whereas the RedFIR system is applicable for the full size of a football pitch.

## 6 Conclusion

We conclude that the RedFIR system (version 1.1) installed indoors in the Test and Application Center L.I.N.K. is able to reliably track the movement of the ball with an RMSE of less than 13cm. This shows the applicability of RedFIR's football tracking for kinematic, tactical and time-motion analyses. Since the ball is the main object of interest in football, knowledge about its movement forms the basis for an automated detection of ball possession, passes and every tactical analysis that involves the ball.

As the system was developed for outdoor applications we expect the systems accuracy to be better than the one presented here, and aim on redoing the tests with a system that is installed outdoors in future work. Since the range of velocities was only between 45km/h and 61km/h we aim on doing a more thorough testing of the accuracy for lower and higher velocities. As a football can typically reach a speed of up to 120km/h it will be interesting to see how the accuracy changes for these high velocities.

## 7 Acknowledgement

This contribution was supported by the Bavarian Ministry of Economic Affairs and Media, Energy and Technology as a part of the Bavarian project 'Leistungszentrum Elektroniksysteme (LZE)'.

## References

1. Bland, J.M., Altman, D.G.: Statistical methods for assessing agreement between measurement. Biochimica Clinica 11, 399–404 (1987)

2. Castellano, J., Alvarez-Pastor, D., Bradley, P.: Evaluation of research using computerised tracking systems (amisco and prozone) to analyse physical performance in elite soccer: a systematic review. Sports Med 44, 701–712 (2014)
3. Chakraborty, B., Meher, S.: A trajectory-based ball detection and tracking system with applications to shooting angle and velocity estimation in basketball videos. In: 2013 Annual IEEE India Conference (INDICON). pp. 1–6. IEEE (2013)
4. Chen, H.T., Tsai, W.J., Lee, S.Y., Yu, J.Y.: Ball tracking and 3d trajectory approximation with applications to tactics analysis from single-camera volleyball sequences. Multimed Tools Appl 60, 641–667 (2011)
5. Choppin, S., Goodwill, S.R., Haake, S.J., Miller, S.: 3d player testing at the wimbledon qualifying tournament. In: Miller, S., Capel-Davies, J. (eds.) Tennis science and technology 3. pp. 333–340. International tennis federation (2007)
6. Di Salvo, V., Collins, A., McNeill, B., Cardinale, M.: Validation of Prozone: A new video-based performance analysis system. Journal of Performance Analysis in Sport 6(1), 108–119 (2006)
7. Eidloth, A., Lehmann, K., Edelhaeusser, T., von der Gruen, T.: The test and application center for localization systems L.I.N.K. In: International Conference on Indoor Positioning and Indoor Navigation. pp. 27–30 (2014)
8. Frencken, W., Lemmink, K., Dellemann, N.: Soccer-specific accuracy and validity of the local position measurement (LPM) system. Journal of Science and Medicine in Sport 13, 641–645 (2010)
9. Gomez, G., Herrera Lopez, P., Link, D., Eskofier, B.: Tracking of ball and players in beach volleyball videos. PLoS ONE 9(11), e111730 (11 2014)
10. von der Grün, T., Franke, N., Wolf, D., Witt, N., Eidloth, A.: A real-time tracking system for football match and training analysis. In: Microelectronic Systems. pp. 199–212. Springer Berlin (2011)
11. Gueziec, A.: Tracking pitches for broadcast television. Computer 35, 38–43 (2002)
12. Kelley, J., Choppin, S., Goodwill, S., Haake, S.: Validation of a live, automatic ball velocity and spin rate finder in tennis. Procedia Engineering 2(2), 2967 – 2972 (2010)
13. Liu, S.X., Jiang, L., Garner, J., Vermette, S.: Video based soccer ball tracking. 2010 IEEE Southwest Symposium on Image Analysis & Interpretation pp. 53–56 (2010)
14. Moeslund, T., Thomas, G., Hilton, A.: Computer Vision in Sports, Advances in Computer Vision and Pattern Recognition. Springer (2015)
15. Mutschler, C., Ziekow, H., Jerzak, Z.: The debs 2013 grand challenge. In: Proceedings of the 7th International Conference on Distributed Event-Based Systems. pp. 283–294 (2013)
16. Ogris, G., Leser, R., Horsak, B., Kornfeind, P., Heller, M., Baca, A.: Accuracy of the LPM tracking system considering dynamic position changes. Journal of Sports Sciences 30(14), 1503–1511 (2012)
17. Owens, N., Harris, C., Stennet, C.: Hawk-eye tennis system. Proc Inf Conf Visual Information Engineering 2003 (2), 182–185 (2003)
18. Siegle, M., Stevens, T., Lames, M.: Design of an accuracy study for position detection in football. Journal of Sports Sciences 31(2), 166–172 (2013)

# Stance Phase Detection for Walking and Running Using an IMU Periodicity-based Approach

Yang Zhao[1], Markus Brahms[2], David Gerhard[1], and John Barden[2]

[1] Department of Computer Science, University of Regina, Regina, SK, Canada
[2] Faculty of Kinesiology and Health Studies, University of Regina,
Regina, SK, Canada

**Abstract.** This paper presents a novel stance phase detection procedure based on observations from a foot-mounted inertial measurement unit (IMU). A frequency-tracking algorithm from the field of audio analysis was applied to the inertial signal to obtain information about gait cycle duration. Afterwards, this information was used to determine the stance phase point for the next gait cycle. This periodicity-based stance-phase detection procedure was found to be superior to traditional threshold-based algorithms, significantly reducing the number of insertion and deletion errors, with less dependence on selected threshold values.

## 1 Introduction

IMUs have successfully been used to solve the problem of localization or Pedestrian Dead Reckoning (PDR) [3, 9, 13, 14]. The general idea is to integrate the recorded accelerometer and gyroscope signals to obtain velocity, displacement and orientation. Since sensor noise accumulates during the integration process, the calculated position may differ substantially from the actual location. When walking or running, this *drift error* may be reduced by periodically performing a zero-update (ZUPT) of velocity during the *stance* phase of each gait cycle.

Most stance phase detection algorithms are based on the idea that the magnitude of the IMU measurement drops below a certain threshold during the stance phase. This threshold may be a static value or a dynamic moving average [1]. Thresholding does not perform well when gait patterns, surfaces, or shoes are variable and/or different from what was used to determine the original threshold(s). Simple thresholding has also been shown to be inconsistent for running [2]. In order to overcome the limitations of traditional threshold approaches, some authors have used adaptive thresholds [11] as well as hidden Markov models [4].

Spectral analysis techniques (Fourier, wavelet etc.) offer alternative ways to analyze gait data and extract fundamental biomechanical variables [8, 12]. Used extensively for audio signal analysis (and other domains), these techniques are able to extract the *fundamental frequency* of human locomotor data ($f_0$), by calculating the reciprocal of the periodicity. Although these techniques have been used to identify individual events (*e.g.*, initial contact and toe-off) within

P. Chung et al. (eds.), *Proceedings of the 10th International Symposium on Computer Science in Sports (ISCSS)*, Advances in Intelligent Systems and Computing 392, DOI 10.1007/978-3-319-24560-7_29

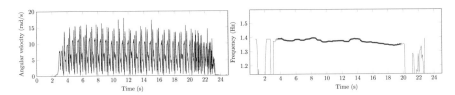

Fig. 1: Gyroscope pitch extraction from jogging. Strongly periodic sections are in blue, with stride rate near 1.4 Hz. Weakly periodic (green) and Non-periodic (magenta) show the initial stance (0 s to 3.6 s) and other movement after 20 s.

a single gait cycle, they have not been applied to solve the problems associated with PDR or stance phase detection.

Based on the theoretical work of using periodic information to determine velocity and position for a general oscillatory movement [6], we present a novel stance phase detection algorithm based on analyzing the periodicity of foot-mounted IMU data. We demonstrate that this approach reduces insertion and deletion errors of stance phases compared to traditional threshold methods.

## 2    Methods

This section describes the development and workflow for our new periodicity-based stance phase detection system.

### 2.1    Pitch extraction from IMU data

Periodicity refers to the rate at which a signal predictably repeats. It is closely correlated to the *fundamental frequency* ($f_0$), which in acoustics research, refers to the lowest frequency component among a set of harmonically related components [10]. The extraction of $f_0$ from an audio signal is known as pitch extraction, and is a well-studied topic. IMU signals relating to repetitive human motion (such as a walk cycle) also possess a similar pattern of periodicity. To extract the "pitch" of inertial gait measurements, we can re-interpret the signal as if it were audio, and apply a time-domain pitch tracking algorithm. For example, the YIN algorithm [5] is a well-known and highly accurate autocorrelation-based pitch tracking method.

Fig. 1 shows the "pitch track" of a 24-second gyroscope magnitude signal using the YIN algorithm. The algorithm marks sections with different colours depending on whether it detects strong periodicity, weak periodicity or aperiodicity. Aperiodicity may indicate the beginning or end of a trial, a pause, or an unusual foot movement. Excluding these sections, the stride rate can be quickly derived from the pitch tracking series, with a greater granularity than peak-to-peak methods traditionally used for stride rate.

### 2.2    Stance phase estimation

The pitch track from Section 2.1 also enables the estimation of stride-to-stride events. For stance phase detection with a gyroscope, prior knowledge tells us that it occurs when the magnitude of the gyroscope signal reaches the lowest

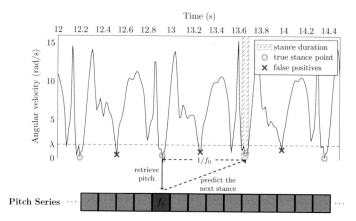

Fig. 2: Periodicity-based stance phase prediction process. The pitch shown in red is from the previous stance phase, and predicts the following stance-phase point.

point of the gait cycle. We can estimate the stance phase of the *next* gait cycle from the current stance position and the extracted pitch signal.

Fig. 2 demonstrates how the pitch series can be used to identify subsequent stance phase points. If $S_p$ is the identified stance point, $h$ is the hop size (*e.g.*, 8 samples), and $\hat{P} = \{f_0^{m+1}, f_0^{m+2}, \ldots, f_0^{m+n}\}$ is the pitched segment of the pitch series $P = \{f_0^1, f_0^2, \ldots, f_0^N\}$; then if $S_p$ is in $\hat{P}$, the frame index $k$ would be:

$$k = [S_p/h] \tag{1}$$

where $[k]$ is the nearest integer to $k$. Since all frames in $\hat{P}$ are pitched, the *k-th* frame is also pitched, thus the position of the next stance point $S_q$ is

$$S_q = S_p + 1/f_0^k \ . \tag{2}$$

This calculation is repeated until the stance point position is greater than that associated with the last pitched frame $(m + n)$ in $\hat{P}$, which equals $h \times (m + n)$.

These predicted stance points can be extended to stance durations with a threshold $\lambda$. The samples to the left and right of each stance point are included in the stance phase until the first sample that exceeds $\lambda$ is met, leading to a stance duration marked as the gray area in Fig. 2.

Note that compared to a traditional simple threshold or peak detection approach, small-value samples can easily be identified as false positives, as depicted with a red cross in Fig. 2. With the periodicity-based approach, these false-positives are automatically skipped.

## 2.3   Initial stance point selection

The algorithm described in Section 2.2 predicts successive stance phases based on an initial stance point. The selection of the initial stance point is critical, because any error will be passed on to the following stance point. Therefore, we select our initial stance point based on the following three criteria. A sample can be an initial stance point only if:

(a) the pitch frame associated with it is considered a periodic frame.

(b) it is the local minimum within the first periodic cycle.

(c) it results in appropriate stance points in subsequent cycles.

During the stance phase, the magnitude of the gyroscope should be at a local minimum. The initial stance point should be chosen as the local minimum when the signal first enters a good periodic state, since any local minima before periodic motion would not be associated with a regular gait cycle. Within one cycle, more than one local minima can occur. Selecting the lowest local minimum is usually sufficient, but occasionally it may lead to an erroneous stance position.

Suppose $S_i$ is the cycle minimum of the $i$-th cycle with a cycle length $l_i$. Criteria (c) states that $S_i$ can only be the initial stance point if:

$$\left| \frac{(S_{i+1} - S_i) - l_i}{l_i} \right| \leq \delta \tag{3}$$

where $\delta$ is the tolerance of error (*i.e.*, 0.1) in seeking the initial stance point. We start with $S_1$ and if $S_1$ fails this condition, we continue to inspect $S_2$, $S_3$, etc.

If the initial stance point that is eventually found is not within the first cycle (for example when the stance point was not a local minimum in the first cycle), a backward version of the algorithm is applied to find all the stance points that were missed in the process of finding the initial stance point.

## 3 Results

In this section, the accuracy of the periodicity-based approach in identifying stance points is investigated and compared with the traditional threshold method. We also consider path tracking with the improved stance phase detection process.

### 3.1 Data acquisition

We chose to evaluate our new method on the task of running because this form of locomotion has received less attention than walking. Additionally, due to the higher locomotor speed, stance phase detection in running is generally considered to be more difficult than for walking [2]. We collected foot-mounted inertial data from 8 participants with an IMU (Xsens, MTw) firmly attached to their right foot. Each subject provided written consent for participation and the data collection was approved by the University ethics committee. Sensor data was recorded at 100 Hz. Additionally, three reflective markers were attached to the foot to track the position of the foot segment with an optical tracking camera system (Vicon, USA). The camera data was processed to determine the ground truth for the timing of stance. Each participant performed twenty linear 10 m runs through the laboratory.

To demonstrate that our method works for pedestrian tracking, we used the ID2 example from a published standard data set [1]. This data, which involves a male participant performing rectangular-shape walking in a laboratory setting for 30 seconds, provides foot-mounted IMU data, motion capture data (for ground truth) and video.

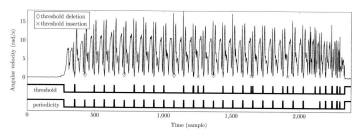

Fig. 3: Stance-phase detection with the traditional threshold method vs periodicity-based method.

## 3.2 Stance phase detection

The traditional threshold-based approach exhibits significant deletion and insertion errors. In [1], a second threshold $\sigma$ was used to filter out false positives. However, the insertion errors can occur at any point between two successive stances, leaving part of the false positives still unresolved. Reducing $\lambda$ provides a more rigorous selection of stance phase and reduces insertion errors, but it can also increase deletion errors. Studies that have used this method weigh one type of error against another and manually pick a threshold for each experiment. For the periodicity-based approach, $\lambda$ is only used for the stance duration, therefore reducing threshold-based errors. Fig. 3 shows an example comparison between the threshold method and the periodicity-based method. The same threshold ($\lambda = 0.7$) is used for both methods. For the traditional method, deletion errors can be clearly seen around sample 430, 710, 1080, and 1370 while insertion errors also occur at sample 1250, 1650, and 1910. The errors are not present in the periodicity-based approach. The section after sample 2100 is aperiodic.

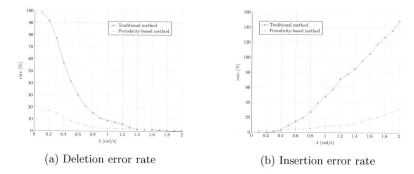

(a) Deletion error rate          (b) Insertion error rate

Fig. 4: Error rate change with a range of different $\lambda$ values on 562 true stances.

A total of 562 stance phases were collected from the 161 running trials. Ground truth for stance phases was acquired using motion-capture. We did not consider the duration of the stance phase, so detected stance was verified when it was in proximity to ground truth. The error rate was calculated as the number of error instances divided by the number of ground truth stances, and is evaluated against a range of typical $\lambda$ values.

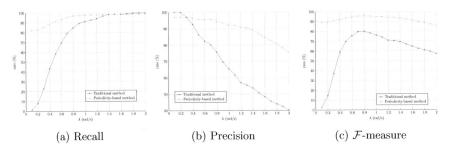

(a) Recall                    (b) Precision                    (c) $\mathcal{F}$-measure

Fig. 5: Stance-phase detection on a range of different $\lambda$ values.

As can be seen from Fig. 4a, A low $\lambda$ results in deletions for both methods, with deletion errors approaching 100% for the traditional method. The periodicity-based approach has some deletions at low $\lambda$ because it downgrades to the traditional method for non-pitched sections. As $\lambda$ increases, the deletion error for both methods decreases. Due to occasional boundary errors and possible pitch detection errors, the periodicity-based method is unable to reach a 0% deletion rate, however, it reaches a deletion rate of 5% when $\lambda = 0.7$ whereas the traditional method requires $\lambda \geq 1.3$ to obtain the same deletion rate.

Fig. 4b compares insertion error rates. When $\lambda$ is low, the traditional method retrieves very few stances, thus the insertion error rate is also low. As $\lambda$ increases, insertion errors rise significantly, and the number of insertions exceeds that of the true stances starting from $\lambda = 1.58$. The periodicity-based method shows very few insertions for low $\lambda$, and insertion errors rise more slowly than the traditional method.

Figs. 5a and 5b show that as $\lambda$ increases, recall increases and precision decreases for both methods, but both recall and precision are significantly higher when using the periodicity-based method. Fig. 5c compares the more balanced $\mathcal{F}$-measure, with the periodicity-based method gaining an $\mathcal{F}$-score of around 90% for all $\lambda$ values, while the $\mathcal{F}$-score for the traditional method is lower, fluctuates more widely, and reaches a maximum of only 80% at $\lambda = 0.8$. Note that this "best" threshold value $\lambda$ varied from experiment to experiment, making the choice of a fixed $\lambda$ for the traditional method problematic.

## 3.3 Pedestrian dead-reckoning

Using ZUPT [7], Fig. 6 shows how the periodicity-based method outperformed the traditional threshold method in the PDR tracking task. Four representative $\lambda$ values were selected. The basic shape of the walking path (*i.e.*, two rectangles with different sizes and an overlap in the lower part) can be recognized with both approaches through all the $\lambda$ values. However, deviation from the true path is also obvious in several situations.

The traditional approach provided a best tracking result when $\lambda = 0.4$. This is also the $\lambda$ value that caused the least deletion and insertion errors. In comparison, the periodicity-based approach had stable tracking across all $\lambda$ values. For each $\lambda$ value, it produced an equal or better tracking path than the traditional approach. For the traditional threshold method, large $\lambda$ caused

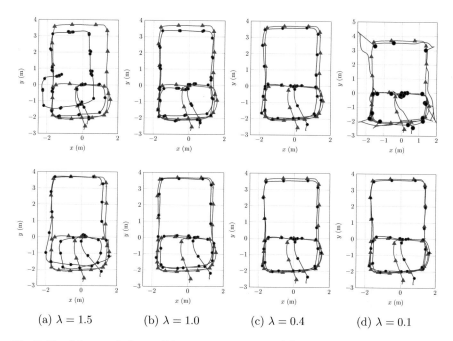

(a) $\lambda = 1.5$    (b) $\lambda = 1.0$    (c) $\lambda = 0.4$    (d) $\lambda = 0.1$

Fig. 6: Tracking path for walking experiment ID2 [1]. Traditional method (top) versus periodicity-based method (bottom), with motion capture ground truth in red, and IMU methods in blue. Markers indicate stance points.

the ZUPT to over-perform, whereas a small $\lambda$ did the opposite. Both the over- and under-performed ZUPTs caused several artifacts in the path tracking. By repairing the deletions, the periodicity-based approach was able to fix all artifacts that deletion errors might have caused. Also, by maximizing the rejection of insertion errors when $\lambda$ is high, the periodicity approach significantly reduced the chances of over-ZUPTs and improved the accuracy of tracking.

## 4  Discussion and Conclusions

This new periodicity-based algorithm is intended not to entirely replace the traditional approach, but to work in tandem with it. When the signal is identified as aperiodic (as in the start or end of a trial, or during a pause), the periodicity-based method downgrades to the traditional threshold approach. As a result, the algorithm can still preserve the performance of the traditional approach when the signal is aperiodic, but it provides far better accuracy for the periodic sections. The method is not dependent on the choice of axis or even the choice of sensor (since any axis of the gyroscope or accelerometer will exhibit similar periodicity). The proposed method may also be used to estimate other stride events as long as a reasonable initial feature classification can be made.

The performance of the periodicity-based algorithm is sensitive to the accu- racy of the initial stance point. Although three rules are used to increase the reliability of finding the starting stance point, it could be executed more fre- quently and more conditions could be added to make the initial detection less prone to error.

The additional pitch tracking process does not bring noticeable computational overload, due to the availability of optimized pitch extractors such as YIN. Practically, we found that the time spent on extracting the pitch for a minute long signal in our data set was almost negligible.

The most notable advantage of this method is that no threshold tuning is needed. The elimination of stance deletion and insertion errors brings significant improvement and convenience to PDR tracking. The technique has been tested on both walking and running on indoor surfaces. Performing similar tests in outdoor environments makes the collection of ground truth problematic, and is therefore left for future work.

# References

1. Angermann, M., Robertson, P., Kemptner, T., Khider, M.: A high precision reference data set for pedestrian navigation using foot-mounted inertial sensors. In: International Conference on Indoor Positioning and Indoor Navigation (IPIN) (2010)
2. Bichler, S., Ogris, G., Kremser, V., Schwab, F., Knott, S., Baca, A.: Towards high-precision IMU/GPS-based stride-parameter determination in an outdoor runners' scenario. Procedia Engineering 34, 592–597 (2012)
3. Brand, T.J., Phillips, R.E.: Foot-to-foot range measurement as an aid to personal navigation. In: Proceedings of the 59th Annual Meeting of The Institute of Navigation and CIGTF 22nd Guidance Test Symposium (2003)
4. Callmer, J., Tornqvist, D., Gustafsson, F.: Probabilistic stand still detection using foot mounted IMU. In: 13th Conference on Information Fusion. pp. 1–7 (2010)
5. De Cheveigné, A., Kawahara, H.: Yin, a fundamental frequency estimator for speech and music. The Journal of the Acoustical Society of America 111(4) (2002)
6. Estrada, A., Efimov, D., Perruquetti, W.: Position and velocity estimation through acceleration measurements. In: 19th IFAC World Congress (2014)
7. Fischer, C., Talkad Sukumar, P., Hazas, M.: Tutorial: Implementing a Pedestrian Tracker Using Inertial Sensors. IEEE Pervasive Computing 12(2), 17–27 (2013)
8. Forsman, P.M., Toppila, E.M., Hæggström, E.O.: Wavelet Analysis to Detect Gait Events. In: 31st Annual International Conference of the IEEE EMBS (2009)
9. Gadeke, T., Schmid, J., Zahnlecker, M., Stork, W., Muller-Glaser, K.D.: Smartphone pedestrian navigation by foot-IMU sensor fusion. 2012 Ubiquitous Positioning, Indoor Navigation, and Location Based Service (UPINLBS) pp. 1–8 (2012)
10. Gerhard, D.: Pitch extraction and fundamental frequency: History and current techniques. Regina: Department of Computer Science, University of Regina (2003)
11. Greene, B.R., McGrath, D., O'Neill, R., O'Donovan, K.J., Burns, A., Caulfield, B.: An adaptive gyroscope-based algorithm for temporal gait analysis. Medical and Biological Engineering and Computing 48(12), 1251–1260 (2010)
12. Han, D., Renaudin, V., Ortiz, M.: Smartphone based gait analysis using STFT and wavelet transform for indoor navigation. In: International Conference on Indoor Positioning and Indoor Navigation (2014)
13. Jiménez, A.R., Seco, F., Prieto, C., Guevara, J.: A comparison of pedestrian dead-reckoning algorithms using a low-cost MEMS IMU. In: 6th IEEE International Symposium on Intelligent Signal Processing. pp. 37–42 (2009)
14. Xu, Z., Wei, J., Zhang, B., Yang, W.: A Robust Method to Detect Zero Velocity for Improved 3D Personal Navigation Using Inertial Sensors. Sensors 15(4) (2015)

# Gamification of Exercise and Fitness using Wearable Activity Trackers

Zhao Zhao, S. Ali Etemad and Ali Arya

Carleton University, 1125 Colonel By Dr, Ottawa, ON K1S 5B6, Cananda

**Abstract.** Wearable technologies are a growing industry with significant potential in different aspects of health and fitness. Gamification of health and fitness, on the other hand, has recently become a popular field of research. Accordingly, we believe that wearable devices have the potential to be utilized towards gamification of fitness and exercise. In this paper, we first review several popular activity tracking wearable devices, their characteristics and specifications, and their application programming interface (API) capabilities and availabilities, which will enable them to be employed by third party developers for the purpose at hand. The feasibility and potential advantages of utilizing wearables for gamification of health and fitness are then discussed. Finally, we develop a pilot prototype as a case-study for this concept, and perform preliminary user studies which will help further explore the proposed concept.

## 1 Introduction

Smart watches, smart glasses, gesture controllers, health monitors, and activity trackers are all part of the emerging landscape of wearable technologies, which are believed to change our lives. The number of wearable devices has seen a significant growth since 2011[3]. Activity and fitness related devices are one of the most dominant categories of wearables and attract a significant amount of research and development [11, 19–21].

On the other hand, gamification, which is the process of utilizing games for tackling particular problems, has become a popular field of research due to increased capabilities and ubiquity of smart phones and PCs [22, 14, 18]. Since exercise and fitness are often physically strenuous, alternative motivators such as entertainment and encouragement through games are considered effective ways for appealing to a wider audience [8–10, 13]. We believe gamification of health and fitness can benefit both novice and professional users alike, and result in short-term engagement as well as long-term improvement through intelligent game-based objectives.

In this paper, we suggest that wearable technologies are suitable platforms as sensing and communication portals for interacting with gamified health and fitness applications and software. In other words, this work proposes and explores the overlap of three different fields, namely digital games and gamification, health and fitness, and wearable technologies. This concept is illustrated in Figure 1. Following a review of activity tracking wearable devices, we present an

© Springer International Publishing Switzerland 2016
P. Chung et al. (eds.), *Proceedings of the 10th International Symposium on Computer Science in Sports (ISCSS)*, Advances in Intelligent Systems and Computing 392, DOI 10.1007/978-3-319-24560-7_30

analysis and discussion on application programming interfaces (API) of existing wearables as APIs play a critical role in order for gamified fitness and health applications to be popularized. Subsequently, as a proof of concept, we design and prototype an exercise game, which utilizes wearable devices for interaction. We perform a preliminary user study, and demonstrate the validity of our proposal.

**Fig. 1.** The three general fields that relate to this work are illustrated. The target field is presented as the overlap of the three fields shown in white

## 2   Related Work

### 2.1   Gamification of Health and Fitness

Gamification is defined as the process of utilizing games or game-like reward and penalty, competition, and goal-based systems in order to increase engagement, incentivize users and popularize particular activities [12]. Lately, the use of gamification in different fields related to exercise, fitness, and health in general, has become common. An evidence for this is the availability of many applications under the category of health and fitness in the application stores of iOS, Android, and Windows, with some aspects of gamification [6].

In the past few years, researchers have also devoted considerable attention to games for exercise, fitness, and health. For instance, Nenonen et al. [17] proposed the use of heart rate as an interaction method for video games. They suggested that heart rate interaction could be utilized in different exercises as users found the concept interesting. In [4], Buttussi and Chittaro proposed a user-adaptive game for jogging, in which they combined the use of a GPS device and a pulse oximeter worn on the user's ear. Their user study results showed that the game motivated users, while having other benefits such as training users to jog as a cardiovascular exercise. Mokka et al. [15] introduced a cycling fitness game that users could play in a virtual environment. Their pilot user test showed that virtual environments could be a motivating factor for exercise. Finally in [5], Campbell et al. focused on everyday fitness games and indicated that for applications that people use in their everyday lives, designs should be fun and sustainable, and adapt to behavioural changes. They applied this idea to the design of a game.

## 2.2   Wearable Technologies

Generally, wearable devices are technological gadgets worn by users, and activity trackers are wearable devices that monitor and record a person's physical fitness activity [7]]. They are fundamentally upgraded versions of pedometers and similar devices [16] and use accelerometers and altimeters to calculate distance and speed, estimate the overall physical activity, calculate calorie expenditure, and in some cases also monitor and graph heart rate and quality of sleep. Some more advanced wearables monitor muscle activity through electromyography, measure body's hydration, estimate lactic acid production, and more [1].

Most of the popular wearable activity trackers provide an API for third party developers. Resource API refers to the access to user resources (read or modify). Subscription API is a set of functions through which the third party applications could be notified when user data changes, which allows applications to have the user's latest data without having to implement a polling or scheduling system to retrieve users' data. Bluetooth API refers to the authorization of third party applications for communicating directly with the device via Bluetooth without the need to synchronize over the web. These three types of API are considered to be critical for developers in order to build applications. Table 1 presents a summary of the availability of different types of API for several wearable devices that can be utilized for the purpose at hand. As we can see, most of the devices provide web-based resource access and subscription API that allow third parties to build applications while Bluetooth APIs and software development kits (SDK) are only provided by few devices. An SDK is generally a collection of tools and functions provided for building new applications on a particular platform[2]. Accordingly, it is generally easier to develop applications for devices that provide an SDK.

## 3   Application Design

To further explore the notion of fitness-related games with wearable devices as interaction mediums, we prototyped two different applications on the iOS platform. The first game is a realistic type of game where users can select either a running or cycling mode. Accordingly, an avatar of a running or cycling subject is depicted on the top half of the screen, while an opponent is depicted on the bottom half. The opponent can either be programmed to complete a course over

**Table 1.** The three general fields that relate to this work are illustrated. The target field is presented as the overlap of the three fields shown in white

| | Resource API | Subscription API | Bluetooth API | Android SDK | iOS SDK |
|---|:---:|:---:|:---:|:---:|:---:|
| Fitbit | ✓ | ✓ | | ✓ | ✓ |
| Nike Fuelband | ✓ | ✓ | ✓ | | ✓ |
| Pebble | ✓ | ✓ | ✓ | ✓ | ✓ |
| Samsung Gear Fit | ✓ | ✓ | ✓ | ✓ | |
| Jawbone Up | ✓ | ✓ | | ✓ | ✓ |
| Garmin Vivofit | ✓ | | | | |
| Misfit Shine | ✓ | ✓ | | | |
| LEO | ✓ | ✓ | ✓ | | |

a fixed period of time (single-player) or correspond to a real user playing in real-time (multi-player). When the multi-player setting is selected, the number of opponents can be entered by the user. Finally, upon completion, the winner is announced, and race metrics are presented in a post-workout summary page. The metrics include duration, distance, maximum speed, average speed, and calorie expenditure among others. Simulated screens from the prototype are presented in Figure 2 (a).

The second game is a more abstract and cartoony application in the form of a goal-based game. In this game, a flowerbed is illustrated, where faster and more consistent exercise on a daily basis results in a more rapid blooming and growth of the flowers. Abandoning the routine for a few days will result in the death of the flowers. Users can unlock higher level and different types and number of flowers when certain goals and milestones in terms of distance, duration of time, speed, calorie expenditure, acceleration, and others are achieved. Similar to the first game, data and summary screens are presented after each session. Figure 2 (b) illustrates simulated screen from this game.

Generally, the first application is a more serious and realistic game, which will potentially engage more professional and athletic audiences, while the second one is more geared towards novice and younger users. While both games enable users to post their results online to social media such as Facebook, the former is mostly expected to leverage athletic competition, while we believe the latter is mostly suitable and attractive for its social network component.

Finally, we mentioned that both games enable two modes: running and cycling. Accordingly, while for running both arm and leg-based wearable devices can be utilized as interaction devices (since there is sufficient arm movement during running), for cycling, an arm or wrist-based device cannot be utilized. This is because on stationary bicycles, the arms/wrists do not have sufficient motion for the device to estimate the overall motion of the body and physical activity. As a result, leg-based wearables will be required for the cycling option of both games.

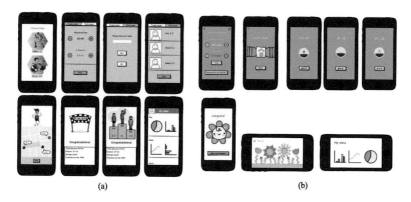

**Fig. 2.** The general design of application prototypes 1 and 2 are presented in (a) and (b) respectively

# 4 User Study and Discussions

## 4.1 User Study

We performed pilot user studies on the mock-up of the two games. Five users were invited to use this app and fill a five-point Likert scale paper-based questionnaire. Three were males and two were females, with the average age of 26 and standard deviation of 2.3 years. In order to evaluate our proposal, three main questions were asked (see Table 2). The questions are meant to evaluate the likelihood of subjects using the games, whether the application motivates subjects to exercise more, and understand the overall satisfaction with the application. The results for this study are presented in Figure 3. The responses were rating values from 1 to 5.

**Table 2.** The major questions used to assess the usefulness of the proposed approach.

| Question 1 | Do you find this kind of application motivating to exercise? |
|---|---|
| Question 2 | How likely are you to use this application again? |
| Question 3 | How would you rate your overall satisfaction with this application? |

## 4.2 Discussions

The number of subjects in the user study was not sufficient to make concrete statistical conclusions. Nonetheless, the study provides some insight as to whether the approach is successful and one that would motivate users or not. According to the results of the study, both games received average ratings of higher or equal to 3/5 on all counts of motivation, engagement, and satisfaction. The

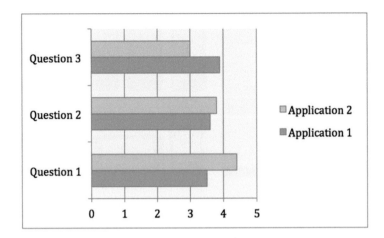

**Fig. 3.** The general design of application prototypes 1 and 2 are presented in (a) and (b) respectively

second game is more likely to motivate users to exercise (4.4/5 compared to 3.5/5 respectively), while also being more engaging (3.8/5 compared to 3.6/5). Interestingly, the first application has a higher overall satisfaction score (3.9/5 compared to 3/5).

The overall results suggest that both player vs. player and goal-based games are feasible approaches to gamification of exercise and fitness, and possess the potential for being utilized in conjunction with wearable devices for this purpose. When further explored the notion of player vs. player as opposed to goal-based games, some subjects suggested that they were in favor goal-based games in order to avoid competitive factors during exercise. These subjects mentioned that the rewards that they would receive when achieving a goal is a sufficient motivator for them to exercise more frequently. On the other hand, some subjects stated that only competition could motivate and engage them to exercise more frequently.

## 5   Future work

Future work includes optimizing the design and implementation of the applications as well as integration of the apps with two wearable devices of different form-factors (for e.g. arm-based and leg-based) will be carried out. The games will be played with multiple subjects, and more details regarding usability will be compiled and statistical conclusions will be drawn. The effect of factors such as leg-based vs. arm-based applications, goal-based vs. multi-player games, animation and graphics, the activity/sport (running, cycling, or etc.), and several others will be studied. Finally, the implications and usefulness of the approach including both short and long-term effects in terms of exercise habits, motivation, and fitness will be studied.

# 6 Conclusion

In this paper we proposed a novel approach for gamification of exercise and fitness, where wearable technologies are utilized for interaction with exercise games. We first reviewed the general status of some activity trackers and their API capabilities and availabilities. We then designed and prototyped a player-vs-player as well as a single-player goal-based game. Finally, pilot user studies on the applications were reported followed by a discussion on the feasibility and potential advantages of utilizing wearables for gamification of health and fitness. Results show that based on existing technologies and user needs, the idea of employing wearables activity trackers for gamification of exercise and fitness is feasible, motivating, and engaging.

# References

1. Gesturelogic inc., home. `http://gesturelogic.com/`, accessed June 10, 2015
2. Techterms.com, sdk (software development kit) definition. `http://techterms.com/definition/sdk/`, accessed June 10, 2015
3. Wearable device market value 2010-2018 forecast.statista. `http://www.statista.com/statistics/259372/wearable-device-marketvalue`, accessed June 7, 2015
4. Buttussi, F., Chittaro, L.: Smarter phones for healthier lifestyles: An adaptive fitness game. Pervasive Computing, IEEE 9(4), 51–57 (2010)
5. Campbell, T., Ngo, B., Fogarty, J.: Game design principles in everyday fitness applications. In: Proceedings of the 2008 ACM conference on Computer supported cooperative work. pp. 249–252. ACM (2008)
6. Deterding, S., Dixon, D., Khaled, R., Nacke, L.: From game design elements to gamefulness: defining gamification. In: Proceedings of the 15th International Academic MindTrek Conference: Envisioning Future Media Environments. pp. 9–15. ACM (2011)
7. Fritz, T., Huang, E.M., Murphy, G.C., Zimmermann, T.: Persuasive technology in the real world: a study of long-term use of activity sensing devices for fitness. In: Proceedings of the SIGCHI Conference on Human Factors in Computing Systems. pp. 487–496. ACM (2014)
8. Hamari, J., Koivisto, J.: Social motivations to use gamification: An empirical study of gamifying exercise. (2013)
9. Hamari, J., Koivisto, J., Sarsa, H.: Does gamification work?–a literature review of empirical studies on gamification. In: System Sciences (HICSS), 2014 47th Hawaii International Conference on. pp. 3025–3034. IEEE (2014)
10. Herger, M.: Gamification in Healthcare and Fitness, vol. 8. EGC Media (2015)
11. Lara, O.D., Labrador, M.A.: A survey on human activity recognition using wearable sensors. Communications Surveys & Tutorials, IEEE 15(3), 1192–1209 (2013)
12. Lister, C., West, J.H., Cannon, B., Sax, T., Brodegard, D.: Just a fad? gamification in health and fitness apps. JMIR serious games 2(2) (2014)

13. Macvean, A., Robertson, J.: Understanding exergame users' physical activity, motivation and behavior over time. In: Proceedings of the SIGCHI Conference on Human Factors in Computing Systems. pp. 1251–1260. ACM (2013)

14. McGonigal, J.: Reality is broken: Why games make us better and how they can change the world. Penguin (2011)

15. Mokka, S., Väätänen, A., Heinilä, J., Välkkynen, P.: Fitness computer game with a bodily user interface. In: Proceedings of the second international conference on Entertainment computing. pp. 1–3. Carnegie Mellon University (2003)

16. Murphy, S.L.: Review of physical activity measurement using accelerometers in older adults: considerations for research design and conduct. Preventive medicine 48(2), 108–114 (2009)

17. Nenonen, V., Lindblad, A., Häkkinen, V., Laitinen, T., Jouhtio, M., Hämäläinen, P.: Using heart rate to control an interactive game. In: Proceedings of the SIGCHI conference on Human factors in computing systems. pp. 853–856. ACM (2007)

18. Nicholson, S.: A user-centered theoretical framework for meaningful gamification. Games+ Learning+ Society 8(1) (2012)

19. Patel, S., Park, H., Bonato, P., Chan, L., Rodgers, M., et al.: A review of wearable sensors and systems with application in rehabilitation. J Neuroeng Rehabil 9(12), 1–17 (2012)

20. Shih, P.C., Han, K., Poole, E.S., Rosson, M.B., Carroll, J.M.: Use and adoption challenges of wearable activity trackers. iConference 2015 Proceedings (2015)

21. Tanenbaum, J., Tanenbaum, K., Isbister, K., Abe, K., Sullivan, A., Anzivino, L.: Costumes and wearables as game controllers. In: Proceedings of the Ninth International Conference on Tangible, Embedded, and Embodied Interaction. pp. 477–480. ACM (2015)

22. Zichermann, G., Cunningham, C.: Gamification by design: Implementing game mechanics in web and mobile apps. " O'Reilly Media, Inc." (2011)

# Part VII
# Neural Cognitive Training

# Training of Spatial Competencies by Means of Gesture-controlled Sports Games

Aleksandra Dominiak[1] and Josef Wiemeyer[2]

Institute of Sport Science, Technische Universitaet Darmstadt
[1] dominiak@gugw.tu-darmstadt.de
[2] wiemeyer@sport.tu-darmstadt.de

**Abstract.** Spatial competencies are one of the most prominent gender differences in view of cognitive abilities. Psychological and interdisciplinary games studies proved a general malleability of spatial competencies. In addition, some studies prove that spatial competencies can be improved by playing First-Person Shooter games. Furthermore studies show that individual spatial experiences are of great importance for the development of spatial competencies. Consequently individual differences in spatial competencies are constituted by learning, training and development factors. Playing video games count to these experiences and are therefore considered as beneficial means of training intervention.

This research project examines suitable training methods for spatial competencies on the basis of gesture-controlled sports games. The aim is to test the effectiveness of this kind of training. It is expected that participants in a 9 hour training intervention will improve their spatial competencies after playing a mimetic Dance game. A small N study shows that significant effects can be expected when studying the optimal sample size.

## 1 Introduction

From a biological perspective all creatures have to interact with their environment in order to survive. The ability to perceive environmental stimuli in order to transform sensory information into motor (re)action, is essential to human and animal organisms. Motor action is always movement in space. Thus spatial activities and competencies are also of great importance in sport science.

Due to an inhomogeneous usage of terminology in current studies concerning spatial competencies it was necessary to construct an own definition and categorization. We consider spatial competencies as an umbrella term for spatial abilities, spatial skills, spatial orientation and spatial visualization. The categorization is partly based on a system used by Uttal et al. [8], which is based on research in linguistics, cognition and neuroscience. Here, two dimensions of spatial abilities are distinguished: an intrinsic vs. extrinsic and a dynamic vs. static dimension. This classification was used to categorize spatial tests in a meta-analysis. For the research project presented in this paper it is used as starting point for an own taxonomy.

© Springer International Publishing Switzerland 2016
P. Chung et al. (eds.), *Proceedings of the 10th International Symposium on Computer Science in Sports (ISCSS)*, Advances in Intelligent Systems and Computing 392, DOI 10.1007/978-3-319-24560-7_31

The criteria of spatial competencies are structured according to the following criteria: specificity, modality, motion state, content and reference system (see Table 1).

Specificity distinguishes abilities and skills. Abilities refer to a generic capability, for example spatial orientation. Skills on the other hand denominate specific operations which can be assigned to different abilities, example way-finding.

Modality refers to the kind of processed information. Information processing includes cognitive, perceptual and sensory-motor modes or levels, as well as a combination of all three modes.

Motion state is either dynamic or static.

The content of spatial information is important as different content seems to be processed by different brain structures. For example, mental rotations of organic bodies and decontextualized objects are based on different neural mechanisms [4]. The content of spatial (re)presentation is described as artificial or biologic. In this regard, artificial denotes all non-biological contents.

The reference system can be intrinsic or extrinsic. Intrinsic processing is used when individuals are generally defining objects and their characteristics. Extrinsic information refers to object constellations and their environment.

**Table 1.** Taxonomy of spatial competencies

**Spatial competencies**

| Criteria | Categories | | |
|---|---|---|---|
| **Specificity** | Skills | Abilities | |
| **Modality** | Perceptual | Cognitive | Sensorimotor |
| **Motion state** | Static | Dynamic | |
| **Content** | Artificial | Biological | |
| **Reference system** | Intrinsic | Extrinsic | |

Based on this taxonomy, the research project presented in this paper aims to answer the question how spatial competencies are acquired and how they can be improved.

In current studies concerning spatial competencies a link between the quality of spatial competencies and STEM field carriers of individuals is assumed [6]. Therefore, in engineering, it is of great importance to identify factors and sources, that influence individual performance in spatial competencies. Different individual spatial experiences accumulated over life span, are assumed to be the reason for performance differences in the STEM field. To assess spatial experiences in the life span, Deno [1] introduced the spatial experience inventory (SEI) with 312 Items. The main predictor variables of the items were Formal Academic Subjects, Non-Academic Activities and Sports in five life span periods: pre-school, elementary school, middle or junior high school, high school and vocational school, and postsecondary. The participants' answers were scored using a four grade scale (1 – 4 points) depending on the frequency of participation for each activity in each time period. SEI scores were related to performance in the Mental Rotations Test (MRT) published by Vandenberg and Kuse [9]. On the one hand, gender-specific experiences were identified that correlated with high MRT performances. For male participants playing with certain toys (Lego bricks, log building sets), for female participants watching educational TV shows in infancy were identified as influencing factors. Irrespective of gender, performance in MRT correlated with the following biographical experiences: repairing activities, playing with building sets, model building, navigating vehicles, using hand and power tools, different masculine connoted sports, such as football or archery and video game play. Sorby, Leopold and Gorska [7] also found a relation between MRT performance and video gaming experience.

Feng, Spance and Pratt [2] show that a 10 hour training using a First-Person-Shooter game lead to higher performances in MRT compared to a non-action video game control group in post-test. Overall action video games are considered as a beneficial training method of spatial competencies [8]. However, in the context of spatial performance and motor activity, the usefulness of gesture-controlled sports games as training method in spatial competencies is still an open question.

## 2 Methods

A small N study was performed to compare the training effects of a mimetic gesture-controlled Dance game to a First-Person-Shooter game and a non-intervention control group on spatial performance. The individual spatial performance of the participants was measured using a testing profile that includes spatial competencies in general, as opposed to one particular spatial skill. The battery includes the redrawn Vandenberg & Kuse MRT by Peters et al. [5], the Hidden-Figures-Test, the Water-Level-Test, and two self-developed Mental Rotations Tests MRT-Bio and MRT-Art (see Figure 1 a-e). The individual spatial experience was assessed using a qualitative guideline interview. The participants' answers were scored using a four-grade scale (0 – 3 points) to build a spatial experience score.

Undergraduates at the Technische Universitaet Darmstadt, Germany (N=12; age range: 20-30 years) participated for course credit or no compensation at all. The optimal sample size was calculated apriori using the power (1-$\beta$=0.94; $\eta^2$=0.39) found in the study by Feng, Spence and Pratt [2], as the optimal sample size would have been N=6 per group, we decided to reduce sample size to only 4 participants per group in the initial stage. All participants reported low current video gaming – at most once a week. They were randomly assigned by lot to one of three groups: Dance, Shooter or Control. Participants assigned to the training groups attended a 9 hours training either playing a shooter (Fable: The Journey) or a dance game (Dance Paradise) on the Xbox 360° Kinect console in the sport science laboratory at Technische Universitaet Darmstadt. Before dividing into groups and starting the training the first session was dedicated to guideline interview and pre-test. Training was delivered in six 1.5 hour sessions over three weeks. Spatial competencies were assessed before training (pre-test), after 4.5 hours of training (mid-test) and after 9 hours of training (post-test). Participants in the control group underwent the testing profile in week one, two and three, as did he training groups. Data were analyzed using a 3 (groups: Dance, Shooter or Control) x 3 (pre-, mid- and post-test) ANOVA with repeated measures on the test factor.

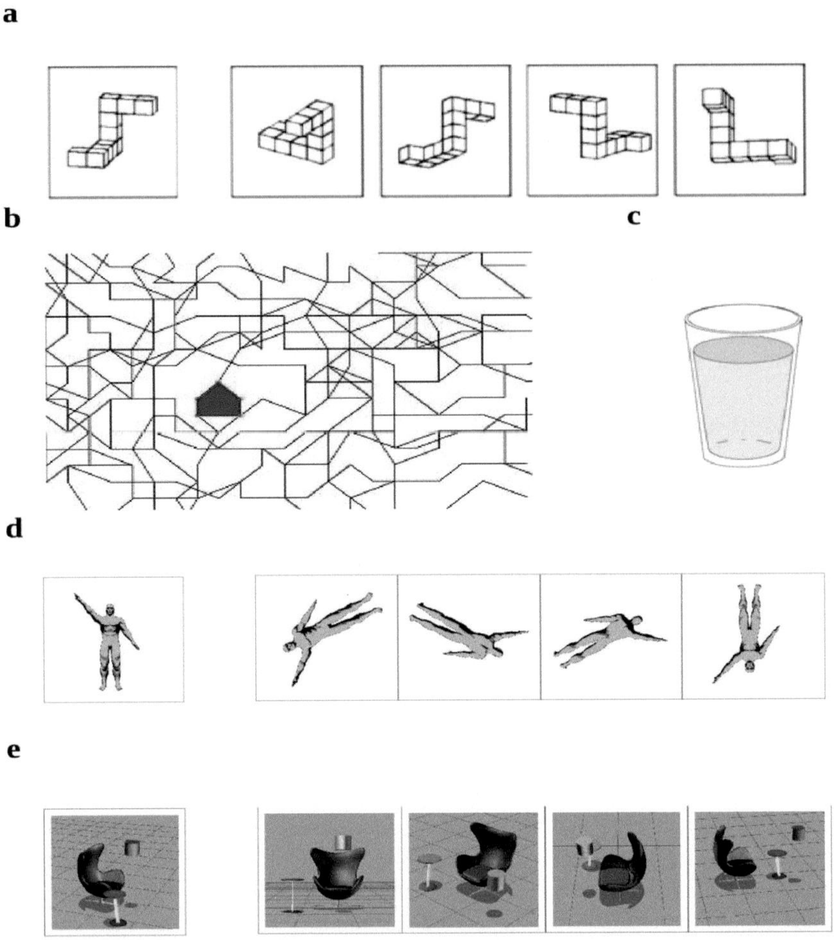

**a**

**b**                                                              **c**

**d**

**e**

**Fig. 1.** a) Sample mental rotation task item **b)** Hidden-Figures-Test item **c)** sample Water-Level-Test item **d)** sample MRT-Bio item **e)** sample MRT-Art item. In a, d, and 3 the task is to identify the correct rotations of the template (left picture). In b the task is to detect the red marked figure hidden in crowded pictures. In c the task is to draw the water levels into glasses tilted in different angles.

# 3    Results and discussion

The full results of the conducted studies are displayed in Table 2. In reference to the three times of measurement MRT-Peters, MRT-Bio and the Hidden-Figures-Test show significant effects. The training groups show no significant training effects compared to the control group. The measured effects are therefore due to test repetition.

**Table 2.** Results of 3x3 ANOVA with factors group and time of measurement (TM)

MRT-Perters

|          | df1   | df2    | F      | p       | $\eta^2$ |
|----------|-------|--------|--------|---------|----------|
| Group    | 2     | 9      | 0.485  | 0.631   |          |
| TM       | 2     | 18     | 22.219 | <0.001* | 0.721    |
| GroupxTM | 4     | 18     | 0.204  | 0.933   |          |

Water-Level-Test

|          | df1   | df2    | F      | p       | $\eta^2$ |
|----------|-------|--------|--------|---------|----------|
| Group    | 2     | 9      | 2.492  | 0.138   |          |
| TM       | 1.118 | 10.066 | 0.68   | 0.445   |          |
| GroupxTM | 2.237 | 10.066 | 0.381  | 0.715   |          |

MRT-Bio

|          | df1   | df2    | F      | p       | $\eta^2$ |
|----------|-------|--------|--------|---------|----------|
| Group    | 2     | 9      | 0.816  | 0.472   |          |
| TM       | 2     | 18     | 3.85   | 0.041*  |          |
| GroupxTM | 4     | 18     | 1.373  | 0.282   |          |

Hidden-Figures-Test

|          | df1   | df2    | F      | p       | $\eta^2$ |
|----------|-------|--------|--------|---------|----------|
| Group    | 2     | 9      | 1.083  | 0.379   |          |
| TM       | 2     | 18     | 11.684 | 0.001*  | 0.565    |
| GroupxTM | 4     | 18     | 2.447  | 0.084   |          |

MRT-Art

|          | df1   | df2    | F      | p       | $\eta^2$ |
|----------|-------|--------|--------|---------|----------|
| Group    | 2     | 9      | 0.685  | 0.529   |          |
| TM       | 1.119 | 10.075 | 0.571  | 0.486   |          |
| GroupxTM | 2.239 | 10.075 | 0.598  | 0.586   |          |

*p<0.05

Figure 2 illustrates the results of the MRT-Peters test in reference to the different training groups. Considering the fact, that this small N-study was underpowered

(Power: $1-\beta = 0.735$), we expect to measure significant effects when the optimal sample size is studied. The optimal sample size is N=6 per group for significant effects. Participants will be assigned to matching triads based on their pre-test results in the testing profile.

The action game will be replaced by Medal of Honor: Pacific Assault on the Xbox 360° console, which has already been used by Feng, Spence and Pratt [2]. The dance training will be delivered using two games Dance Paradise and Dance Central after two sessions of introduction. Afterwards, the participants are free to choose between the two games. This procedure considers the participant's reports, that one game only becomes boring after a while. Spatial experience will be evaluated using a questionnaire to build more exact spatial experience scores.

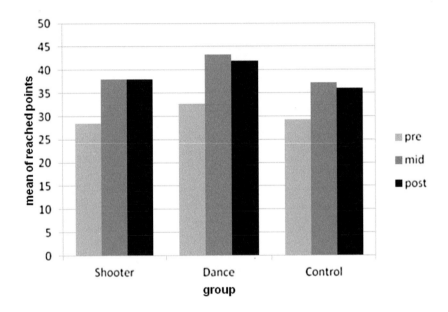

**Fig. 2.** MRT-Peters chart

## 4    Conclusion

The results of this small N study are as expected. Due to the small number of participants we chose in this initial stage of the project, we are aware of the fact, that the significance of the results could be lower. This trial was nonetheless important for study design testing due to the long-term character of this study. We consider the study design acceptable.

We expect to find significant training effects for both video game trainings compared to the non-intervention control group at least in Peters et al. MRT. In addition, we hope that the participants with small spatial experience scores will benefit more from the video game training. An effective training of spatial competencies by means of gesture-controlled sports games would give especially women the opportunity to easily benefit from video gaming considering that they mostly prefer this kind of games, because of their less time intensive character [3].

# References

1. Deno, J. A.: The Relationship of previous Experiences to Spatial Visualization Ability. Engineering Design Graphics Journal, 59 (3), 5–17 (1995).
2. Feng, J., Spence, I. & Pratt, J.: Playing an Action Video Game Reduces Gender Differences in Spatial Cognition. Psychological Science, 18 (10), 850–855 (2007)
3. Hartmann, T., & Klimmt, C.: Gender and computer games: Exploring females' dislikes. Journal of Computer-Mediated Communication, 11 (2), 910 – 931 (2006)
4. Kosslyn, S.M., Digirolamo, G.J., Thompson, W.L., & Alpert, N.M.: Mental rotation of objects versus hands: Neural mechanisms revealed by positron emission tomography. Psychophysiology, 35, 151 – 161 (1998)
5. Peters, M., Laeng, B., Latham, K., Jackson, M., Zaiyouna, R. & Richardson, C.: A Redrawn Vandenberg and Kuse Mental Rotations Test - Different Versions and Factors That Affect Performance. Brain and Cognition, 28 (1), 39–58 (1995)
6. Sorby, S. A. & Baartmans, B. J.: The Development and Assessment of a Course for Enhancing the 3-D Spatial Visualization Skills of First Year Engineering Students. Journal of Engineering Education, 301–307 (2000)
7. Sorby, S. A., Leopold, C. & Gorska, R.: Cross-Cultural Comparisons of Gender Differences in the Spatial Skills of Engineering Students. Journal of Women and Minorities in Science and Engineering (5), 279–291 (1999)
8. Uttal, D. H., Meadow, N. G., Tipton, E., Hand, L. L., Alden, A. R., Warren, C. et al.: The Malleability of Spatial Skills: A Meta-Analysis of Training Studies. Psychological Bulletin (2), 352 – 402 (2012)
9. Vandenberg, S. G.; Kuse, A. R.: Mental Rotations, a Group Test of Three-Dimensional Spatial Visualization. Perceptual and Motor Skills, 47 (2), 599–604 (1978)

# Methods to Assess Mental Rotation and Motor Imagery

Melanie Dietz[1] & Josef Wiemeyer[2]

[1] Research Training Program Topology of Technology, Technische Universität Darmstadt
dietz@sport.tu-darmstadt.de
[2] Institute of Sport Science, Technische Universität Darmstadt
wiemeyer@sport.tu-darmstadt.de

**Abstract.** Spatial abilities are the basic precondition to process visual-spatial information no matter whether that information derives from a digital or a real source. They play an important role in many different areas of life. Vocation, leisure, education and sport are affected by perceptional abilities. However, in sport science this field of research is quite underestimated. Mental rotation tasks occur for example during the early phase of motor learning processes when a motor imagery as an internal representation develops to govern the imitation of the presented model. While IT-based applications are applied to support the imagery process, their actual impact on learning may depend on the level of spatial abilities. To study the relations between spatial abilities and the development of a proper motor imagery two different test instruments, MRT BIO (Mental Rotation Test for biological objects) and PiCaST (Picture Card Selection Test), were developed. Those testing tools and the associated research options are presented in the current paper.

**Keywords:** Motor Imagery, motor representation, motor learning, motor learning control, spatial abilities, mental rotation

## 1    Introduction

Spatial abilities play an important role in nearly every sphere of human life. Beside their role as a key component in human perception and action spatial abilities enable the processing of visual-spatial information [15]. Especially in technical professions, like in STEM fields, spatial abilities have an important relevance and seem to be a performance limiting factor. Spatial abilities can be improved by appropriate training [16]. However, there is no generally agreed definition of spatial abilities [18]. This is partly due to the heterogeneous operationalisations of spatial skills including perceptual and manipulative tests. According Linn and Peterson [6, p. 1482] the term "generally refers to skill in representing, transforming, generating, and recalling symbolic, nonlinguistic information". A common feature of many definitions is the ability to represent and interact with space according to the task-specific demands.

The relevance of spatial abilities for sport science arises from the fact that visual-spatial components are considered to be a key to transfer performance [11]. During the early phase of motor learning, learners establish an internal representation with

© Springer International Publishing Switzerland 2016
P. Chung et al. (eds.), *Proceedings of the 10th International Symposium on Computer Science in Sports (ISCSS)*, Advances in Intelligent Systems and Computing 392, DOI 10.1007/978-3-319-24560-7_32

251

strong emphasis on visual-spatial aspects. This motor imagery seems to be important for the success when learning new movements. As an example, we presume that a high level of spatial abilities fosters the acquisition of appropriate motor imagery from movements presented by video sources.

The purpose of this paper is to discuss two methods for assessing mental rotation (one specific spatial ability) and motor imagery. The theoretical background is analyzed for both abilities. Based on existing assessment methods, we will describe newly developed methods and discuss the respective research options.

## 2    Mental Rotation

"The ability to mentally turn an object in space and be able to then mentally rotate a different object in the same way" represents one out of five different spatial abilities regarding the distinction of McGee [7 quoted by 14, p. 280]. Due to the fact that rotations in mind are part of many practices during motor learning (e.g. imitation learning) our focus is set on this spatial ability. Our research concerns questions about the relations between mental rotation and motor learning processes.

### 2.1    Assessing Mental Rotation

According to Hahn [5] the first mental rotation test (MRT) was developed by Shepard and Metzler in the early 1970s. Within this first MRT it is the test person's task to compare two cube objects regarding their congruence. Based on the first object it is either possible or not to rotate the second object in a way that it matches with the first one. The test person has to decide whether this is possible or not. Following this test design the most popular MRT was developed by Vandenberg and Kuse in 1978 and redeveloped by Peters in 1995. Each task consists of five pictures: one template, two correct rotations and two distractors. Similar to the former test it is the test person's task to identify the correct rotations of the template [17]. Figure 1 shows one task of the redeveloped test. Compared to the template, shown in the first place, pictures number one and three are congruent. The other pictures (number two and four) are distractors and represent a rotation of the mirrored template.

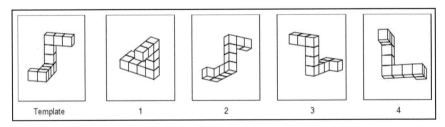

**Fig. 1.** One task of the MRT from Peters following the original MRT from Vandenberg and Kuse [13, p. 42]. The task of the test person is to identify the correct rotations of the template (see text for further details).

Another line of development is shown within the MRT of Parsons [12]. In this test, instead of artificial cube objects, human body postures are used to test mental rotation. Contrary to the other MRTs, there is only one rotated item per task and the test person has to make a left-right judgment of body parts and decide whether the left or right arm is outstretched. Figure 2 shows one task from the Parsons test where the left arm is outstretched.

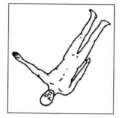

**Fig. 2.** One task of Parsons' MRT with the outstretched left arm [12, p. 174].

## 2.2    MRT BIO

Referring to the two presented MRTs we developed a new testing tool that applies the stimuli of Parsons [12] and the test logic of Peters [13]. This new test instrument is called MRT BIO and exists in one basic paper and pencil version as well as in four different computer-based versions. All different test versions consist of six different items and are limited to a testing time of 90 seconds. The digital test instruments differ from each other regarding the time constraint on one hand which is either strict (15 seconds per task) or variable (90 seconds for 6 tasks). On the other hand the order of the items is either constant or random within multiple testings. In principle the digital test versions are expected to improve controlling the test situation and to generate extra information, i.e. recording decisions and changes of decisions. Furthermore the test result is automatically calculated and displayed in the output file. However, a reliability analyses with those different test instruments did not confirm the expected advantages [2]. Only the paper and pencil version showed a satisfactory test-retest reliability of r = 0.861 (N = 50). Therefore this version should be used within future research. Figure 3 shows one exemplary task of MRT BIO.

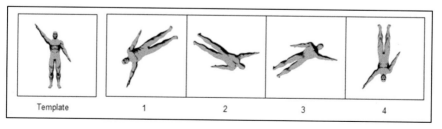

**Fig. 3.** One item of the newly developed MRT BIO. The test combines MRT Peters and MRT Parsons. The task of the test person is to identify the correct rotations of the template. Compared to the template, picture one and three represent the correct rotations.

## 3    Motor Imagery

Imagery in general is defined as "a perception-like process in the absence of any external stimulus input" [9, p. 307]. Motor imagery can be considered as an (at least partly) "conscious explicit process operating on representations of movements" [19, p. 38]. Motor imagery is a cognitive correlate of motor learning comprising prescriptive and interpretative knowledge i.e., knowledge serving the preparation, execution and evaluation of movements [19].

Generally, every stimulus modality is imaginable but regarding motor imagery, visual and kinesthetic modalities are of particular interest. Therefore motor imagery refers to specific characteristics of movement.

Figure 4 shows an overview of the possible perspectives and the different actor roles for the relevant modalities of motor imagery.

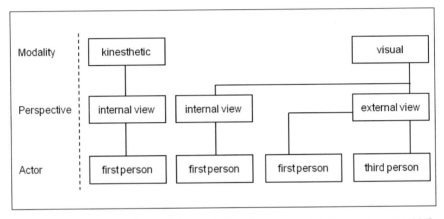

**Fig. 4.** Modalities, Perspectives and Actors of Motor Imagery [modified according 10, p. 221].

Within the early stage of motor learning processes a motor imagery evolves. This prescriptive motor knowledge is initially imprecise due to the fact that the learner does not have extensive experiences regarding the movement to learn. Early motor imagery includes mainly spatial aspects and a rough structure guiding the execution. Due to missing feedback and evaluation of performance an elaborate reference concept does not yet exist [19]. At that time, the expertise of motor imagery has a crucial influence on motor execution and learning success [3, 4]. Müller [8] has also shown that a structured motor imagery benefits early learning. In this learning phase especially the visual-spatial components of motor imagery are of particular importance [11].

Therefore it is essential to use visual-spatial information sources within the early phase of motor learning processes to support learners developing a proper motor imagery.

## 3.1   Assessing Motor Imagery

Due to the fact that motor imagery is a prescriptive and interpretative motor knowledge, it should be possible to assess this construct. Basically the motor imagery can be recorded directly because essential parts are visually coded. For this purpose picture selection tests can be used to arrange items into the correct order to analyze the quality of the developed motor imagery [19].

To assess motor imagery Wiemeyer and Angert [20] developed a computer-aided picture selection test (CAST). Before the test a gymnastic movement sequence is presented to generate a motor imagery. Then the associated CAST is to be completed. In this the correct items have to be selected from an item bundle and arranged into the right sequence [20]. Using this digital test environment enables gathering plenty additional information, i.e. time of picture selection and picture positioning, time of corrections as well as the automatically calculated result. Figure 5 shows a screenshot of the computer-aided picture selection test (CAST). The appropriate pictures have to be selected from the lower part of the program window and put into the selection area where the pictures should be arranged into the correct sequence.

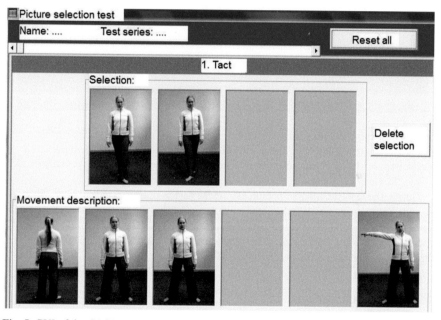

**Fig. 5.** GUI of the CAST. Upper part: target fields; lower part: selection fields. The task of the test person is to select the pictures from the lower part and to arrange them into the correct sequence to the target fields in the upper part.

Due to the fact, that the images of CAST are of a poor quality, it is planned to revise this testing tool. In the meantime an alternative paper-based test instrument (PiCaST) will be used.

## 3.2    PiCaST

Referring to the existing CAST, a new less complex testing tool was developed. The picture card selection test (PiCaST) is based on a motion animation and the associated fixed-images of the motion's key points. The stimuli were created with Poser 9. The animation shows ten different postures that are independent from each other. After each posture a neutral position is shown. According to the optimal number of presentation repetitions for visuomotor learning [1] the animation is shown three times. Directly after the third presentation the test person receives ten stacks with four pictures each. Those pictures can be assigned to four different categories:

1. the correct posture
2. the mirrored posture in the sagittal plane
3. a similar posture with an implemented parameter error
4. the mirrored posture in the sagittal plane with the parameter error

Figure 6 shows the first key point of the animation presented in PiCaST.

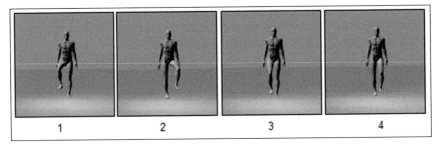

**Fig. 6.** Pictures of the first key point in PiCaST. 1 - correct posture; 2 - mirrored posture; 3 - a similar posture with an implemented parameter error; 4 - mirrored posture plus parameter error

During the test run, the test person can revise earlier decisions at any time. There is no time limit, so the test is terminated voluntarily by the test person. PiCaST exists in two different versions that can be distinguished from each other regarding the observation perspective. One version shows the scene from the front (PiCaST-F – see also figure 5) and the other from behind (PiCaST-B).

## 4    Research options

Using the developed test instruments opens up many research options.

First of all it allows studying differences and advantages of diverse perspectives. Therefore it is possible to answer the question if it is more appropriate to watch a training video from the front or back perspective to give learners a good precondition during the early phase of learning a new movement.

Second, it is possible to compare digital and paper-and-pencil versions of the tests. In this regard and according to our previous results, the expected superiority of the paper-and-pencil version can be tested.

Furthermore the test instruments enable research about the relation between the ability to mentally rotate and the ability to develop a motor imagery based on the repeated observation of the motion animation. Probably this relation is moderated by differential spatial experiences accumulated in the individual biography.

Finally, it is possible to study compatibility/incompatibility effects, i.e., the relevance of the ability to mentally rotate for the development of a motor imagery with a rotation task integrated into the animation. High compatibility is characterized by observing the animation and completing the PiCaST from equal perspectives, e.g. front perspective, whereas low compatibility is characterized by differing perspectives, e.g., observing from the front and displaying the test items from the back and vice versa. Based on the results of the planned studies it will be possible to provide practical advices regarding the importance of spatial abilities in motor learning processes.

# References

1. Daugs, R., Blischke, K., Olivier, N. & Marschall, F. (1989). *Beiträge zum visuomotorischen Lernen im Sport. [Contributions to visual motor learning in sport]* Schorndorf: Hofmann.
2. Dietz M., Dominiak, A. & Wiemeyer, J. (2015). Computer-based methods to assess spatial abilities. In: A. Baca & M. Stöckl (Ed.), *Sportinformatik X - Jahrestagung der dvs-Sektion Sportinformatik vom 10.-12. September 2014 in Wien. (p. 93-99)*. Hamburg: Feldhaus.
3. Fleishman, E. A., & Rich, S. (1963). Role of kinesthetic and spatial-visual abilities in perceptual-motor learning. *Journal of Experimental Psychology, 66* (1), 6.
4. Gregg, M., Hall, C. & Butler, A. (2010). The MIQ-RS: A Suitable Option for Examining Movement Imagery Ability: *eCAM, 7* (2), 249-257.
5. Hahn, N. (2010). *Mentale Rotation bei Vorschulkindern: Geschlechtsunterschiede in der Lateralisierung.* Dissertation, Heinrich-Heine-Universität, Düsseldorf. November, 24th 2014: http://docserv.uni-duesseldorf.de/servlets/DerivateServlet/Derivate-17865/Diss_final _Nicky%20%28pdfA%29.pdf
6. Linn, M. C., & Petersen, A. C. (1985). Emergence and characterization of sex differences in spatial ability: A meta-analysis. *Child development*, 1479-1498.
7. McGee, M.G. (1979). *Human spatial abilities: Sources of sex differences.* New York: Praeger.
8. Müller, H. (1995). *Kognition und motorisches Lernen.[Cognition and motor learning]* Bonn: Holos-Verlag.
9. Munzert, J., Lorey, B. & Zentgraf, K. (2009). Cognitive motor processes: The role of motor imagery in the study of motor representation. *Brain Research Reviews, 60*, 306-326.
10. Munzert, J. & Reiser, M. (2003). Vorstellung und Mentales Training. In: H. Mechling & J. Munzert (Ed.), *Handbuch Bewegungswissenschaft, Bewegungslehre (p. 219-230).* Schorndorf: Hofmann.
11. Panzer, S., Krueger, M., Muehlbauer, T., Kovacs, A.J. &, Shea, C.H. (2009). Inter-manual transfer and prctice: Coding of simple motor sequences. *Acta Psychologica, 131,* 99-109.

12. Parsons, L. M. (1987). Imagined Spatial Transformation of One's Body. *Journal of Experimental Psychology: General, 116* (2), 172-191.
13. Peters, M., Laeng, B., Latham, K., Jackson, M., Zaiyouna, R. & Richardson, C. (1995). A Redrawn Vandenberg and Kuse Mental Rotations Test: Different Versions and Factors That Affect Performance. *Brain and Cognition, 28*, 39-58.
14. Sorby, A., Leopold, C. & Górska, R. (1999). Cross-cultural comparison of gender differences in the spatial skills of engineering students. *Journal of Women and Minorities in Science and Engineering, 5*, 279-291.
15. Souvignier, E. (2000). Förderung räumlicher Fähigkeiten: Trainingsstudien mit lernbeeinträchtigten Schülern. [Enhancing spatial abilities: training studies on pupils with learning difficulties] Münster; New York; München; Berlin: Waxmann.
16. Uttal, D. H., Meadow, N. G., Tipton, E., Hand, L. L., Alden, A. R., Warren, C., & Newcombe, N. S. (2013). The malleability of spatial skills: a meta-analysis of training studies. *Psychological Bulletin, 139* (2), 352-402.
17. Vandenberg, S. G. & Kuse, A. R. (1978). Mental Rotations, a Group Test of Three-Dimensional Spatial Vizualization. *Perceptual and Motor Skills, 47*, 599-604.
18. Voyer, D., Voyer, S. & Bryden, M.P. (1995). Magnitude of Sex Differences in Spatial Abilities: A Meta-Analysis and Consideration of Critical Variables. *Psychological Bulletin, 117* (2), 250-270.
19. Wiemeyer, J. (2001). Conscious representations of movement: structure and assessment. *Motor control and learning in sport science, 1*, 1-12.
20. Wiemeyer, J. & Angert, R. (2011). Computer Methods to Assess Motor Imagery. *International Journal of Computer Science in Sport, 10* (2) 37-53.

# Self-regulated multimedia learning in Sport Science

## Concepts and a field study

Josef Wiemeyer[1] and Bernhard Schmitz[2]

[1] Institute of Sport Science, Technische Universitaet Darmstadt
wiemeyer@sport.tu-darmstadt.de
[2] Institute of Psychology, Technische Universitaet Darmstadt
schmitz@psychologie.tu-darmstadt.de

**Abstract.** To fully exploit the potentials of e-learning, learners have to take more responsibility for their own learning. According to models of self-regulated learning (SRL), learners have to actively engage in planning, execution and reflection of their learning activities. The purpose of this paper is to present a study on the effects of an SRL training on multimedia learning in sport science.

Twenty students attending a course in biomechanics of sport were randomly assigned to a self-regulation training (SRT) or a no-training (NO-SRT) group. The SRT group underwent an SRL training. Before and after the learning period of 2 weeks, the students' knowledge and self-confidence was tested. Although the SRT group spent more time with the program than the NO-SRT group, there was no learning benefit. Test performance of both groups significantly increased from pretest to posttest.

**Keywords:** Self-Regulated Learning, Multimedia, E-Learning, Biomechanics, Sport Science

## 1    Introduction

Multimedia technology has much to offer for enhancing learning compared to traditional teaching methods (e.g., [10]). Learning can become more flexible, more authentic, more communicative, and more motivating for students. However, empirical research and practical experience clearly show that in order to systematically exploit these potentials, a number of moderators have to be taken into consideration. One of these factors is self-regulated learning (SRL). Multimedia learning requires students to self-regulate their learning: Students have to plan, perform, and evaluate their learning activities on their own. Therefore success in multimedia learning is supposed to depend considerably on the ability to self-regulate. The rationale of this paper is that multimedia learning in sport science can be supported by a specific SRL training program. In sport science education the practical application of knowledge to problem solving (i.e., planning a PE lesson or correcting movements) is an important issue. SRL has proven to support problem solving (for a review, [6]).

© Springer International Publishing Switzerland 2016
P. Chung et al. (eds.), *Proceedings of the 10th International Symposium on Computer Science in Sports (ISCSS)*, Advances in Intelligent Systems and Computing 392, DOI 10.1007/978-3-319-24560-7_33

First, we deal with the concept of SRL in more detail. Then we present an experimental field study we performed in order to test the impact of SRL training (SRT) on multimedia learning using the example of biomechanical principles in sport.

## 2    Self-Regulated Learning – Concept, Components, and Models

Whereas the concept of SRL has been discussed in pedagogy for a long time, educational psychology has discovered this construct in the late 1980s [13]. The term 'self-regulation' 'refers to self-generated thoughts, feelings, and actions that are planned and cyclically adapted to the attainment of personal goals' ([12], p.14). SRL has emerged as 'an important new construct in education' ([1], p.445), because of new challenges in modern society such as the necessity for life-long learning in dynamically changing professional fields, and the appearance of new information and communication technologies. According to [9], SRL is a very demanding kind of learning requiring learners to analyze tasks, set goals, know and adopt task-specific learning strategies, reflect their own learning process, evaluate their learning progress, and, if necessary, make corrections to enhance their own learning. Students self-regulating their learning are actively involved in selected or all phases of the learning process.

Regarding the components of SRL, one can distinguish at least two perspectives (for a review, [6]):

— Phase-based approach
   In this approach, different phases of SRL are distinguished, e.g., pre-action, action, and post-action. One of the most popular approach is the three-phase model [12] involving a forethought, performance or volitional control and self-reflection phase. During these phases, characteristic sub-processes take place. Another model has been proposed by [8]. This model distinguishes a preaction, action, and postaction phase. Many tenets of the model could be confirmed by applying a combined diary and time-series approach. Based on this model, [8] developed a training program in order to support SRL. The main focus of the program is on goal setting, learning strategies, and volitional strategies. In particular, the following aspects are emphasized: goal setting, planning and time management, avoidance of procrastination, attention control, self-motivation, and dealing with distraction.
— Activity-based approach
   The numerous learning activities in SRL are usually subdivided into three categories (e.g., [1][9][11]): cognitive, meta-cognitive, and motivational. [12] conceptualizes SRL as a triadic interaction of (covert) personal, behavioural, and environmental processes. We differentiate two main categories: information-related and incentive-related activities. These categories can be divided into further sub-categories: cognitive, meta-cognitive, motoric, motivational, emotional, and volitional.

In conclusion, the application of training programs enhances SRL. First, a direct effect on SRL components is intended and confirmed by the respective studies. Second, primary effects on SRL components should yield an indirect effect on learning outcomes. These effects are also expected in the present study.

# 3    A study on SLR training in sport science education

In sport science education the application of knowledge to solving concrete problems is important. Therefore, in biomechanics education a blended learning concept was implemented consisting of two main parts: First the basic biomechanical knowledge is taught, followed by a second part where the knowledge has to be applied to the analysis of sport movements. In the first part, concepts of 'blended learning' are applied, i.e. a combination of online and offline education. By including SRL in the online learning phase, enhanced learning and transfer was expected.

## 3.1    Methods

**Participants.**
This study consisted of 20 students (5 female, 15 male) of sport science. All students attended a compulsory course in movement science (title: 'Introduction to Sport Biomechanics')[1]. Learning about the biomechanical principles was mandatory. Ten students were randomly selected to constitute the experimental group. The other students formed the waiting control group.

**Multimedia learning environment.**
The students used the multimedia learning environment 'BioPrinz', Version 3.2[2]. This learning environment has been developed by the first author using Macromedia Authorware. The learning environment was distributed by CD-ROM.

The environment comprised two main parts: principles according to [4] and [5] and general information (references, glossary, and links).

The first main part consisted of six chapters:

— Chapter 1.0 gave an overview of the content. The concept of biomechanical principles was explained and the respective principles were enumerated.
— Chapters 1.1 to 1.5 addressed the following biomechanical principles according to [4] and [5]: initial force, optimal trajectory of acceleration, coordination of partial momentum, 'actio – reactio' (Newton's third law of motion), and conservation of momentum. These chapters consisted of seven sections: overview, examples, animations and simulations, definitions and explanations, tasks, exercises, and quiz.

Students could navigate through the environment without any restrictions. In addition, they could use search functions (search by phrases, by chapter, or visited pages). Further options were a print function, an annotation function (supported by a data

---

[1]  Originally 33 students attended the course. The final sample ($n=20$) represents those students who attended both the pretest and posttest. For analysis of log file data, 25 complete data sets were available.

[2]  'BioPrinz' has been awarded the 'Best E-teaching Award 2004' of Darmstadt University of Technology for the 'excellent multimedia processing of dynamic movements'. In the meantime it has moved to the internet (URL: http://bioprinz.ifs-tud.de/).

base), and a help menu. The program could be installed on the computer or run on CD. The program generated a log file whenever the learning environment is used.

**SRL training intervention.**
The SRL training consisted of two sessions, each session lasting 90 minutes. The structure of these sessions is illustrated in Table 1. The training involved five components: goal setting, meta-cognition, time-management, external and internal resources. The components were first introduced theoretically and then exercised practically.

**Table 1.** Structure of the SRL training sessions

| First session | Second session (one week later) |
|---|---|
| 1. Introduction | 1. Reflection: positive and negative experiences |
| 2. SRL model | 2. Repetition: SLR model and components |
| 3. Goal setting | 3. Allocation of external resources |
| 4. Meta-cognition | 4. Allocation of internal resources (self-motivation, volition) |
| 5. Time-management | |
| 6. Example | 5. Game-like repetition |
| 7. Home work | 6. Support of transfer |

**Dependent measures.**
We analysed two dependent measures: learning effort and learning outcome. Learning effort was assessed by two variables: duration and frequency of use. Duration of use was calculated based on the log file data by summing up the durations of all sessions. After each week a separate measure was assessed reflecting initial and later effects. Frequency of use was determined by counting the number of sessions. Learning outcomes were measured by a paper-and-pencil knowledge test. This test consisted of four parts (42 items): knowledge and application of biomechanical concepts and biomechanical principles. The test was administered immediately before and after the two-week learning period. Total score was 58 points. In addition to their solutions of the tasks, the students were asked to document how confident they felt with their answer on a five-point scale. Maximum confidence score was 210 points.

**Design and procedure.**
The experiment was performed according to a 2 (treatment: SRL training vs. no training) × 2 (test: pretest, posttest) two-factor repeated measures design (Table 2).

In the first biomechanics lesson, students were first tested using the paper-and-pencil test mentioned above. Then they were randomly assigned to a training (SRT; n = 9) and a no-training group (NO-SRT; n = 11). All students were delivered a CD-ROM copy of the multimedia learning environment. They were asked to use the CD-ROM to prepare for the next lesson. The duration of the three biomechanics lesson was 90 minutes. The following biomechanics lesson dealt with the biomechanical principles. After the definition of the principles, examples were analysed to apply the principles to sport movements. One week later, i.e., in the third biomechanics lesson, the posttest was performed. Before the second and third lesson, the two SRL trainings session took place for the SRT group.

**Table 2.** Design of the experiment

| Group | Biomechanics lesson 1 (90 min) | SRL lesson 1 (90 min) | Biomechanics lesson 2 (90 min) | SRL lesson 2 (90 min) | Biomechanics lesson 3 (90 min) |
|---|---|---|---|---|---|
| SRT | Randomisation Pretest CD-ROM | SRL training | Biomechanical principles – definitions & examples | SRL training | Posttest |
| NO-SRT | | --- | | --- | |

## 4 Results

### 4.1 Learning effort

Fig. 1 illustrates the results for duration of use. Both groups showed a similar profile: duration of use increased considerably from first to second week (SRT group: 632%, $n = 7$, NO-SRT group: 465%, $n = 10$). A 2 (first and second week) × 2 (experimental group) ANOVA revealed a significant main effect of week, $F(1,15) = 35.83$, $p < .001$, $\eta^2 = 0.71$. The main effect of experimental group was not significant, $F(1,15) = 3.34$, $p = .088$, $\eta^2 = 0.18$. Interaction of experimental group × week was just above the 5% level of significance, $F(1,15) = 4.28$, $p = .056$, $\eta^2 = .22$.

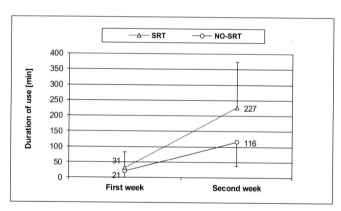

**Fig. 1.** Duration of use (SRT – SRL training group, $n = 7$, NO-SRT – no-training group, $n = 10$)

Frequency of use showed a similar pattern. Frequency of use increased from 1.2 to 2.7 times per week (SRT: from 0.86 to 2.86, NO-SRT: from 1.5 to 2.6). Again, a 2 × 2 ANOVA revealed a significant effect of week, $F(1,15) = 17.00$, $p = .001$, $\eta^2 = 0.53$. The main effect of experimental group was not significant, $F(1,15) = 0.14$, $p = .72$, nor was the interaction, $F(1,15) = 1.43$, $p = .25$.

## 4.2    Learning outcome

Fig. 2 illustrates that both groups improved test scores from pretest to posttest. The respective main effects were significant for test scores, $F(1,18) = 94.56, p < .001, \eta^2 = .84$, and test confidence, $F(1,18) = 199.68, p < .001, \eta^2 = .92$. There was no main effect for experimental groups, for test scores, $F(1,18) = 1.57, p = .23$, and test confidence, $F(1,18) = 0.10, p = .76$. The interaction of experimental group × test was also not significant, test scores, $F(1,18) = 0.11, p = .74$, and test confidence, $F(1,18) = 0.24, p = .63$.

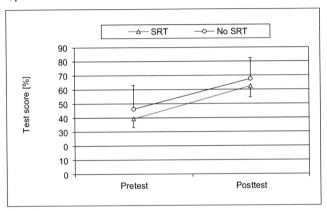

**Fig. 2.** Test scores of the experimental groups (SRT group: $n = 9$; *NO-SRT* group: $n = 11$)

Separate analysis of the learning gains, i.e., differences of posttest and pretest performance, for the four parts of the test, also revealed neither a main effect of experimental groups, $F(1,18) = 0.11, p = .74$, nor an interaction of experimental group × part, $F(3,54) = 1.03, p = .39$.

Furthermore, there was no significant correlation between frequency or duration of use on the one hand and test scores or test confidence on the other hand.

## 5    Discussion

The purpose of the present study was to enhance self-regulated multimedia learning of biomechanical principles by applying a specific model-based SRL training program. We performed a two-factor repeated-measures design with a training and a test factor. The SRL training program consisted of five components (goal setting, time management, meta-cognition, allocation of internal and external resources). Two 90-minute SRL training sessions were performed. We expected that both learning effort and learning outcome was positively affected by the SRL training program.

The results show that the SRL training intervention did not result in significant differences. There was only a tendency that the students of the SRL group spent more time working with the multimedia learning environment during the second week (delayed effect). An increase of learning effort as a direct intervention effect was found in the respective studies of [8]. In the present study, however, SRL training, characterised by training goal setting, meta-cognition, time management, and allocation of internal and external resources, influenced the learning effort only marginally.

Furthermore, the treatment did not result in enhanced learning outcomes of the SRT group compared to the group without SRL training. Therefore, learning efficiency (i.e., ratio of learning effort and learning outcome) was lower in the SRT group. This result is, in a way, consistent with expectations of critics of multimedia learning [10] arguing that a possible increase in learning outcomes may be over-compensated by an increase of learning effort.

The results of the present study may be due to several factors. First, we did not study the 'pure' effect of multimedia learning but rather the effect of 'blended learning'. Because the students discussed the biomechanical principles during the second lesson of the course, possible differences may have been compensated by this additional learning activity. Thus, SRL training and classroom learning were confounded, particularly concerning the application of knowledge to problem solving. Second, the learning period of two weeks may have been too short. Looking at the learning outcomes, only 60 to 70% of the maximum score was achieved. Therefore, a ceiling effect can be excluded. Third, the sample was very heterogeneous. This is, for instance, indicated by the great range of learning duration, ranging from 14 to 552 minutes, and learning gains, ranging from 4 to 24.5 points. Fourth, the study was substantially under-powered (Power: 0.12). A sample of 230 participants is required to detect an effect size of $\eta^2 = 0.006$ for learning.

As an interesting additional result, students not only improved their knowledge of biomechanical principles by using the multimedia program and attending the biomechanics lesson, but they also improved basic biomechanical knowledge. Thus, by dealing with biomechanical problems, which necessarily involves basic biomechanical concepts and the respective calculations (e.g., angular momentum as a product of moment of inertia and angular velocity), a 'side-effect' could be verified. This finding is consistent with results of our previous research using earlier versions of 'BioPrinz'.

## 6 Conclusions

In general, a five-component SRL training consisting of two training sessions had no significant impact on learning effort and learning outcome of students using the multimedia learning environment 'BioPrinz'. However, there was a statistical tendency of an interaction indicating a delayed effect on learning effort, i.e., it became evident in the second week of SRL training. This effect can be considered as a tendency of a direct effect that is consistent with the SRL model proposed by [8]. On the other hand, the missing effect on learning outcome clearly shows that enhanced learning effort does not necessarily influence learning results. In other words, spending more time with multimedia learning does not necessarily enhance learning. This missing direct relationship of (quantitative) learning effort and learning outcome is also reflected by the absence of correlations. Rather, many other factors contribute to learning which may not have been affected differentially by the SRL training.

Future research should isolate the impact of SRL training on multimedia learning by avoiding confounding with classroom instruction. The SRL training program involved five components. Therefore, appropriate instruments should be applied to assess the differential influence of the SRL training program on goal setting, time management, meta-cognitions, and allocation of internal and external resources. This research strategy may improve our knowledge in two ways: First, more detailed information is gained concerning the effects of the training program. Second, also more detailed information is obtained concerning the relationship of the five components of learning effort and the learning outcomes.

# 7    References

1.  Boekaerts, M.: Self-regulated learning: Where we are today. International Journal of Educational Research 31, 445-457 (1999)
2.  Bund, A.: Selbstkontrolle und Bewegungslernen [Self-regulation and motor learning]. Wissenschaftliche Buchgesellschaft, Darmstadt (2008)
3.  Cleary, T. J., Zimmerman, B. J., Keating, T.: Training physical education students to self-regulate during basketball free throw practice. Research Quarterly for Exercise and Sport 77 (2), 251-262 (2006)
4.  Hochmuth, G.: Biomechanik sportlicher Bewegungen (1. ed.). [Biomechanics of sport movements]. Limpert, Frankfurt/ M. (1967)
5.  Hochmuth, G.: Biomechanik sportlicher Bewegungen (5. ed.). [Biomechanics of sport movements]. Sportverlag, Berlin (1982)
6.  Jaekel, F. & Schreiber, C.: Introspection in problem solving. Journal of Problem Solving 6 (1), 20-33 (2013).
7.  Puustinen, M., Pulkkinen, L.: Models of self-regulated learning: A review. Scandinavian Journal of Educational Research 45 (3), 269-283 (2001)
8.  Schmitz, B., Wiese, B. S.: New perspectives for the evaluation of training sessions in self-regulated learning: Time-series analyses of diary data. Contemporary Educational Psychology 31, 64-96 (2006)
9.  Spoerer, N., Brunstein, J. C.: Erfassung selbstregulierten Lernens mit Selbstberichtsverfahren. Ein Überblick zum Stand der Forschung. [Assessing self-regulated learning with self-report measures: A state-of-the-art review]. Zeitschrift für pädagogische Psychologie 20 (3), 147-160 (2006)
10. Wiemeyer, J.: Multimedia in sport – Between illusion and realism. In P. Dabnichki & A. Baca (Eds.), Computers in Sport, pp. 293-317. WIT Press, Southampton, UK (2008)
11. Wolters, C. A.: Regulation of motivation: Evaluating an underemphasized aspect of self-regulated learning. Educational Psychologist 38 (4), 189-205 (2003)
12. Zimmerman, B. J.: Attaining self-regulation. A social cognitive perspective. In M. Boekaerts, P.R. Pintrich, & M. Zeidner (Eds.), Handbook of self-regulation, pp.13-39. Academic Press, San Diego, CA (2000)
13. Zimmerman, B. J., Schunk, D. H. Self-regulated learning and academic achievement: Theory, research, and practice. Springer, New York (1989)

# Author Index

© Springer International Publishing Switzerland 2016
P. Chung et al. (eds.), *Proceedings of the 10th International Symposium
on Computer Science in Sports (ISCSS)*, Advances in Intelligent Systems
and Computing 392, DOI 10.1007/978-3-319-24560-7

Printed in the United States
By Bookmasters